# Nutrient digestion and utilization in farm animals

# Nutrient digestion and utilization in farm animals

*Editor*

**Almiro Salazar**

scitus
academics

**Nutrient digestion and utilization in farm animals**

Edited by **Almiro Salazar**

ISBN: 978-1-68117-625-3
Library of Congress Control Number: 2016947445


SCITUS Academics LLC,
616, Corporate Way, Suite 2, 4766,
Valley Cottage, NY 10989

www.scitusacademics.com



Printed in United States of America on Acid Free Paper

# Preface

Digestion in animals involves mechanical, enzymatic and microbial processes in the gastrointestinal tract that convert large feed particles to a size that animals can absorb, transport and implement. The digestive systems of all farm animals are not exactly the same, so that they are less able and competitive to adapt to a wide variety of free feed resources. Based on the intestinal tract 's structure and functions, farm animals are divided into two main groups: polygastrics or ruminants, for example, cattle, buffaloes, sheep and goats; monogastrics and non-ruminants, rabbits, donkeys, for example, cats, horses, and dogs. The key difference between these two animal groups lies in the stomach structure. Ruminants have three additional stomach compartments in which digestion occurs in different environments instead of non-ruminants. In general, prehension processes for farm animals include digestion, ingestion, grinding or possibly mastication, feed digestion, nutrient absorption and excretion of waste products.

This monograph includes comprehensive information on nutrient digestion and use in poultry, fish cattle, sheep, and pigs' production. The book covers a variety of topics and approaches to animal production and feed assessment. This book will appeal to students, practitioners, and researchers and, for all those dealing with nutrition modeling.

# Table of Contents


Chapter 1. The Use of Probiotics in Animal Nutrition 1

Chapter 2. Comparative performance of broiler chickens offered ten equivalent diets based on three grain sorghum varieties as determined by response surface mixture design 31

Chapter 3. Nutrition and Health of Dairy Animals 67

Chapter 4. Effect of feeding different levels of palm kernel cake fermented by *Paenibacillus polymyxa* ATCC 842 on nutrient digestibility, intestinal morphology, and gut microflora in broiler chickens 117

Chapter 5. Impact of mycotoxin on immune response and consequences for pig health 143

Chapter 6. Animal Models for Study of Diabetes Mellitus 165

Chapter 7. Selenium Biochemistry and Bioavailability: Implications for Animal Agriculture 203

Chapter 8. The Occurrence and Toxicity of Indospicine to Grazing Animals 221

Chapter 9. Effects of spray-dried animal plasma on food intake and apparent nutrient digestibility by cats when added to a wet pet food recipe 243

Chapter 10. Dietary fiber and protein fermentation in the intestine of swine and their interactive effects on gut health and on the environment: A review 265

**Chapter 11.** **Comparative effects of dietary carbohydrases without or with protease on the ileal digestibility of energy and amino acids and $AME_n$ in young broilers** **291**

**Index** **317**

# CHAPTER 1

# The Use of Probiotics in Animal Nutrition

**Hiruta Yirga**

Haramaya University, School of Animal and Range Science,
Dire Dawa, Ethiopia

## ABSTRACT

This paper reviewed issues regarding the use of probiotics (feed additive) in animal feeding. Probiotics are increasingly used in commercial animal production operations to advantageously alter gastrointestinal flora, thereby improving animal health and productivity. The major outcomes from using probiotics include improvement in growth, reduction in mortality, and improvement in feed conversion efficiency. Although it is not well defined, the mechanisms of probiotics activity to achieve their role include alteration in intestinal flora, enhancement of growth of nonpathogenic bacteria, forming lactic acid and hydrogen peroxide, suppression of growth of intestinal pathogens, and enhancement of digestion and utilization of nutrients.

Different probiotics contain different microorganisms which may behave differently, as probiotics are not single entities. It has been indicated that multi-strain preparation is highly efficient in animal feeding. Finally, the efficacy of probiotics have been found to be variable depending on survival rate and stabilities of strains, doses, frequency of administration, interactions with some medicines, health and nutritional status of the animal and the effect of age, stress and genetics of animals.

## KEYWORDS

Feed additive; Probiotics

## 1. INTRODUCTION

In intensive production systems, the nutritional requirements of livestock can be met through supplementation of the limiting nutrients in concentrated form, so that they can produce large quantities of products rapidly. However, the use of feedstuffs that are rapidly fermented in the rumen can create conditions in the rumen that are suboptimal for the fibrolytic microorganisms, thereby impairing fiber digestion [1]. Moreover, nutritional quality of a feed is not only influenced by nutrient content but also by many other aspects such as, hygiene, content of anti-nutritional factors, digestibility, palatability and effect on intestinal health. Hence, the use of feed additives has been an important part of achieving this success [2].

Feed additives are materials that are used to enhance the effectiveness of nutrients and exert their effects in the gut or on the gut wall cells to the animal [3]. They are used for the purpose of promoting animal growth through their effect in increasing feed quality and palatability [2]. Besides, as they are mixed with feeds in non-therapeutic quantities and protecting the animal against all sorts of harmful environmental stresses. Low levels of additives in animal feed contribute to increase production of animal protein for human consumption thereby decrease the cost of animal products [4]. There are a number of feed additives that are used in animal feeds such as antibiotics, probiotics, oligosaccharides, enzymes and organic acids. The use of additives, make end products to be more homogenous and of better quality.

From the different additives, antibiotics have been widely used in livestock diets during the past several decades due to their therapeutic effects [5]. Antibiotics have been used to reduce the frequency of diarrhea under certain conditions and resulted in improvement in performance parameters like body weight gain (BWG) or feed conversion ratio (FCR). These beneficial effects of feed antibiotics are generally explained by modifications of the intestinal bacteria and their interaction with the host animal, including bacterial interactions with intestinal tissue as well as the immune system. Thus, the intestinal micro biota is not only involved in nutrient conversion along the gastrointestinal tract, but may also affect or support animal health [6].

Because of the concern that the use of antibiotics as feed additive might contribute to an increase of bacterial antibiotic resistance, the use of some types of antibiotics have been restricted by some countries beginning from 1970's. Further, European Union (EU) has introduc ed a total ban on the application of antibiotics as feed additives from 2006 onwards [5,7]. Consequently, the change in the consumer's demand for a safe food production coupled with the regulatory issues about the ban of antibiotic growth promoters have ensured a search for natural strategies to modulate gut development and health [6,8]. Hence, many activities were initiated to establish alternative strategies aimed to preventing the growth of pathogenic bacteria in farm animals, to maintain their health and performance. At the same time as making gains in production and efficiency, the industry has had to maximize the health and well-being of the animals and minimize the impact of the industry on the environment [2]. Accordingly, probiotics, prebiotics, organic acids, herbs and essential oils have been suggested as alternatives to antibiotics [6]. From the available alternative, there is presently an increased interest in using natural feed additives such as enzymes, bacterial direct-fed microbials and yeast to help maintain optimum ruminal digestion of feed [1].

Probiotics are live microbial feed supplements which beneficially affect the host by improving its intestinal microbial balance [9]. The application of probiotics provides a potential alternative strategy to the traditional practice of sub therapeutic antibiotic use. In relation to probiotics, several studies observed the beneficial effects on animals including growth enhancement and disease prevention [5]. However, there is still a need to clarify their effectiveness and the underlying mechanisms through which they function.

Feed additives may not be put on the market unless authorization has been given following a scientific evaluation demonstrating that the additive has no harmful effects, on human and animal health and on the environment. Authorizations are granted for specific animal species, specific conditions of use and for ten years periods. Any additives used in feed must be approved for use and then used as directed with respect to inclusion levels and duration of feeding [2]. Correspondingly, in feed regulation, probiotics are included in the group of feed additives for stabilising the microbial communities of the digestive tract in both monogastric and ruminant animals [9].

In this paper, some aspects of probiotics will be discussed. From the many issues, the historical backgrounds, the most common types, the effects exerted on animals and how they function will be highlighted.

# 2. ALTERNATIVES TO ANTIBIOTIC GROWTH PROMOTERS

Considering the intention of organizations and the EU to end all use of antibiotics as growth promoters by 2006, the need for novel strategies to modulate the gastrointestinal environment and microflora metabolism became of top priority [8]. Essentially, the option used to reduce antibiotic use in animals is the development of alternatives to antibiotics that work via similar mechanisms, promoting growth whilst enhancing the FCE. There are many options suggested as an alternative to antibiotic use in animals and some of them are discussed below.

It must be noted that, in considering phasing out or banning antibiotic growth promoters, the quality of any alternatives that could be developed or that are available illegally, must be assessed.

## 2.1 In-feed enzymes

In-feed enzymes are produced as fermentation products from fungi and bacteria and seem to only have a positive effect on the animal. They are routinely added to pig and poultry feeds and work by helping to break down certain components of the feed, such as glucans, proteins and phytates, that the animal may have problems digesting. In-feed enzymes are very effective at maximising FCE and have few drawbacks [7].

## 2.2 Competitive exclusion products

Competitive exclusion products are in-feed microbes consisting of a variety of species of bacteria that are marketed as being "friendly". The mechanism of action is believed to be that, by allowing such bacteria to colonise the gastrointestinal tract. This is the competitive exclusion principle [7]. It is not known how effective the treatment is but it is believed to reduce diarrhea and reduce levels of mortality.

## 2.3 Probiotics

Probiotics are similar to competitive exclusion products. They are believed to improve the overall health of an animal by improving the microbial balance in its gut. The problem with probiotics is the lack of evidence as to their mechanism of action and of the effects on host animals. Probiotics are effective in certain cases, notably in newborn animals or those that have been treated with antibiotics, where they have the same effect as competitive exclusion products. They may also be useful in helping to boost weight gain and feed conversion rates.

## 2.4 Infection control measures

The use of antimicrobials as growth promoting agents rests on their role in controlling infection in growing animals. Similarly, many of the alternatives are aimed at controlling infection, often indirectly. The "specific pathogen-free" system is used to prevent pigs from acquiring many of the diseases that require antibiotic intervention, especially respiratory disease. Vaccination is also used to offer protection against certain pathogens.

## 2.5 Combo strategies

Although different approaches have shown beneficial influences in modulating the fermentation process within the gastrointestinal tract when supplemented alone, evidence is growing for the efficacy of an intriguing new approach. It seems that a combination of more than one novel approach may lead to an even more favorable equilibrium of intestinal metabolism and thus animal welfare and performance. Literature concerning this strategy is still weak, even though some trials have been carried out. This approach takes into account all the different aspects of the gastrointestinal tract: microbiology, nutrient metabolism and tissue requirements [8].

Pro+pre-biotic=symbiotic: The combination of a probiotic and a prebiotic can be a synergistic strategy that beneficially affects the host by improving the survival and the implantation of a direct-fed microbial in the gastrointestinal tract, and by electively stimulating the growth and/or by activating the metabolism of a limited number of health-promoting bacteria [10]. The beneficial response can be more evident when animals are challenged by pathogens or chemicals.

Prebiotic+gut nutrient: As the intestine represents a complex environment, trying to promote the intestinal ecosystem may be best achieved through manipulation of nutrient availability and microbial activity. Following this concept, application of probiotic cultures, alone or in combination with prebiotic oligosaccharides, has been found to ameliorate microbial population patterns in the gastrointestinal tract and, in so doing, favorably affect the host [11]. There have also been a few reports [8,12] about the development of flavorings and herbal extracts for stimulating appetite, as well as for displaying antagonism toward undesirable microbes and improving the antioxidant status of the host and, in so doing, beneficially affecting the health status in swine or poultry.

## 3. PROBIOTICS

Although the digestive tract of all animals is sterile at birth, contact with the mother and the environment leads to the establishment of a varied microflora [3]. There are hundreds of bacterial strains that inhabit both human and animal gastrointestinal tracts. These bacteria include harmful or toxic bacteria that colonize within the digestive tract and produce toxic waste products which lead to gas or bloating, diarrhea, constipation, ulcers or more serious events like food poisoning, and beneficial bacteria. Thus, offering the possibility to exert a positive and completely natural effect on health, well-being and performance of the animal through its autochtonous microflora [9]. Yet, the beneficial microorganisms produce enzymes that complement the digestive ability of the host, and their presence provides a barrier against invading pathogens [3].

As a result of legislation that prohibit the use of sub therapeutic levels of antibacterial, the use of probiotics as a possible alternative to antibiotics has received renewed interest [13]. Successful alternatives to antibiotics are likely to be most effective if they function as a controlling or stabilizing influence on the flora of the gut [14]. Digestive upsets are com mon at times of stress (e.g. weaning), in such conditions, in contrast to the use of antibiotics as nutritional modifiers, which destroy the desirable bacteria as well as the harmful species, the inclusion of probiotics in foods is preferable. Probiotics are designed to encourage certain strains of bacteria in the gut at the expense of less desirable ones [3]. It is believed that gut bacteria have requirements for specific nutrients that may not be adequately provided by the animal's diet. Therefore, feeding these nutrients may promote the growth of gut bacteria, thereby improving the microbial profile in the gut [15]. Presently, there is increasing interest concerning the use of probiotics in the livestock industry.

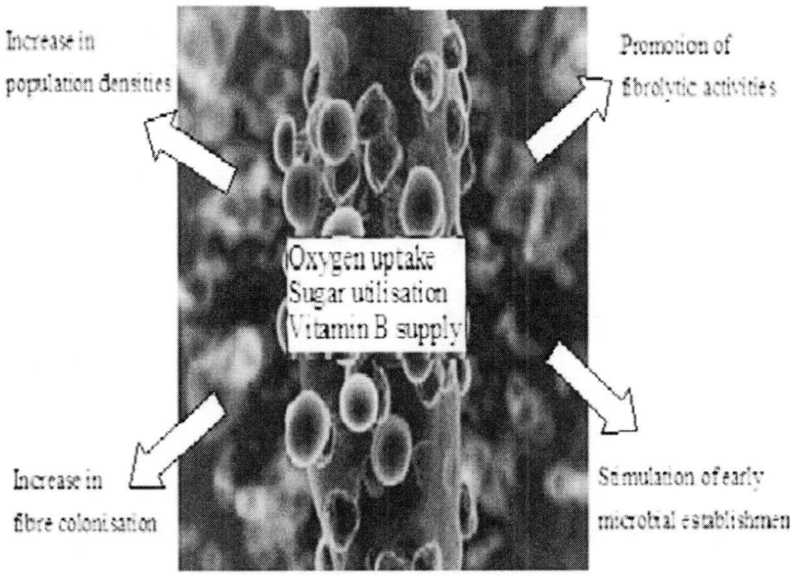


**Figure 1.** Main effects and mechanisms of action of probiotics on rumen microbes (fibre-degrading communities) [16].

## 3.1 Historical notes about probiotics

Historically, the intake of food with microbiological activity began as early as the start of human civilization, with fermented milks likely being the first foods containing active microorganisms. However, the scientific base of the beneficial effects derived by the consumption of fermented lactic products for human use began at the beginning of the 20th century [9]. In 1907, a Russian biologist postulated that the bacteria involved in yoghurt fermentation, Lactobacillus bulgaricus and Streptococcus thermophilus, suppress the putrefactive type fermentations of the intestinal flora and that consumption of these played a role in maintaining health [13,17]. Later, different studies demonstrated the relevant role of the intestinal microflora in the mechanisms of local and systemic defence in front of certain pathogens. Only at the end of the century, it became clear that intestinal microflora had several functions, including metabolic, trophic and protective ones [18]. The health benefits derived from the consumption of foods containing Lactobacillus acidophilus, Bifidobacterium and L. casei are now well documented [2]. In the 1960's and 1970s they were rediscovered for human and animal nutrition.

The value of these new products consisting of live microorganisms was based on the hypothesis that the intake of high levels of certain specific bacteria, with no negative effects on host, could somehow reduce the capacity for the pathogenic bacteria to colonise up to undesirable levels in the digestive tract of the host [17]. The first potent products for animal nutrition to fulfill the specific requirements for feed additives did not appear on the market until the mid-1980's.

## 3.2 Modes of action of probiotics

It was assumed that the effect of probiotics was linked to the gastrointestinal tract and effects on incidence of diarrhea and other gut infections were expected. However, recent work in several different countries has indicated that the effects may be more general [19]. The reason for diverge mechanisms may be due to the different types of probiotics [5]. Probiotics are believed to improve the overall health of an animal by improving the microbial balance in its gut. In general, the mode of action of probiotic feed additives is mainly based on competitive exclusion, bacterial antagonism, and immune modulation [7, 20] .

Competitive exclusion: Competitive exclusion is defined by the ability of normal microflora to protect against the harmful establishment of pathogens. The concept of competitive exclusion indicates that cultures of selected, beneficial microorganisms, supplemented to the feed, compete with potentially harmful bacteria in terms of adhesion sites and organic substrates (mainly carbon and energy sources) [20]. The adhesion to the digestive tract wall could be for different purposes: to prevent colonization by pathogenic microorganisms or to compete for nutrients.

Adhesion to the digestive tract wall to prevent colonization by pathogenic microorganisms: Detrimental bacteria need to become attached to the gut wall to exert their harmful effects [3]. Therefore, an expected effect of the addition of probiotics to the gastrointestinal tract is an increase in normal microflora colonization with inhibition of the adhesion of harmful pathogens on the intestinal epithelium [5], thereby blocking receptor sites and preventing the attachment of other bacteria including harmful species. By doing so, the probiotic bacteria exclude pathogens and thus prevent them from causing infection [7]. The mechanism of colonization is suggested to be associated with certain species within the microflora which can influence the expression of glycol conjugates on epithelial cells that may serve as receptors for the adhesion of bacteria [21].

Different studies have shown the potential of probiotics to decrease the risk of infections and intestinal disorders. Hillman, et al. showed that the growth of E. coli was successfully inhibited by different strains of Lactobacilli. Moreover, it has been reported that, a combination of different lactic acid bacteria significantly reduced the levels of Salmonella in caecal contents of broilers which had been orally inoculated with the pathogen. In addition, in piglets, attachment of E. coli to the small intestinal epithelium has been reported to be inhibited by dietary supplementation with Enterococcus faecium.

Competing with pathogenic bacteria for nutrients in the gut: Probiotics may compete for nutrients and absorption sites with pathogenic bacteria. In addition, competition for energy and nutrients between probiotic and other bacteria may result in a suppression of pathogenic species. The gut is such a rich source of nutrients that it may seem unlikely that microorganisms could not find sufficient food for growth. Probiotics possess a high fermentative activity and stimulate digestion. Lactobacilli are known to produce lactic acid and proteolytic enzymes which can enhance nutrient digestion in the gastrointestinal tract [22]. Different studies demonstrated that probiotics maximized crude protein and energy digestibility compared with those in non-probiotic treatments [22]. However, it should be noted be that the environment only has to be deficient in one essential nutrient in order to inhibit microbial growth. In addition, the ability to rapidly utilize an energy source may reduce the log phase of bacterial growth and make it impossible for the organism to resist the flushing effect exerted by peristalsis [5].

Bacterial antagonism: Probiotic microorganisms, once established in the gut, may produce substances with bactericidal or bacteriostatic properties [7, 20].

Bactericidal activity: Lactobacilli ferment lactose to lactic acid, thereby reducing the pH to a level that harmful bacteria cannot tolerate. Hydrogen peroxide is also produced, which inhibits the growth of Gram-negative bacteria [3]. These substances have a detrimental impact on harmful bacteria, which is primarily due to a lowering of the gut pH. A decrease in pH may partially offset the low secretion of hydrochloric acid in the stomach of weanling piglets. Moreover, live yeasts ferment sugars derived from the degradation of starch, thus competing with the lactic-acid-producing bacteria, and thereby stabilize rumen pH and reduce the risk of acidosis [3]. Improvement in early digestion and intake is brought about by alterations in the numbers and species of microorganisms in the rumen [3].

Neutralization of enterotoxins produced by pathogenic bacteria that cause fluid loss: Probiotic bacteria produce a variety of substances that include organic acids, antioxidants and bacteriocins [3]. These compounds may reduce not only the number of viable pathogenic organisms but may also affect bacterial metabolism and toxin production. Bacteriocins produced by lactic acid bacteria have been reported to be able to permeate the outer membrane of gram-negative bacteria and subsequently induce the inactivation of gram-negative bacteria in conjunction with other enhancing anti-microbial environmental factors such as low temperatures, organic acids and detergents [23].

In addition, they can prevent amine synthesis. Coliform bacteria decarboxylate amino acids to produce amines, which irritate the gut, are toxic and are concurrent with the incidence of diarrhea. If desirable bacteria prevent the coliforms proliferating, then amine production will also be prevented [3].

Immune modulation: Probiotics act as a stimulus for the immune system. Though, analysis and research into the ability of probiotics to influence the immune system of animals and humans is a recent development. According to Lan et al. microbial communities can support the animal's defence against invading pathogens by stimulating gastrointestinal immune response. This may aid the development of the immune system by stimulation of the production of antibodies and increased phagocytic activity [3]. As the immune system is engaged following exposure to probiotic bacteria, any hostile bacteria are also noticed, following increased surveillance by leukocytes, and thus potential pathogens are eliminated [7]. Some probiotic strains such as Lactobacillus have proven to be capable of stimulating the immune system. Fuller explained the immune system to be stimulated in two ways. They can either migrate through the gut wall as viable cells or multiply to a limited extent or antigens released by the dead organisms can be absorbed and stimulate the immune system directly. It is the product of this change which induces the immune response. And currently, it appears to be some relationship between the ability of a strain to translocate and the ability to be immunogenic [19].

However, it is difficult to completely conclude that probiotics contribute significantly to the immune system of the host as they are not intended to eradicate invasive pathogens in the gastrointestinal tract. Therefore, such observed improvements or positive effects are always somewhat compromised due to the animals immune system status and the various applied situations [5]. In addition to the above discussed, other postulated effects include beneficial interaction with bile salts, increased digestive enzyme production, more efficient absorption of nutrients, and greater vitamin production. Several mechanisms have been proposed to explain the effects of probiotics and it is likely that the positive results reported in the different animal studies are due to a combination of some, if not all, of these [3].

## 3.3 Probiotics in animal feeding

Modern rearing methods which include unnatural rearing conditions and diets induce stress and can cause changes in the composition of the microflora which compromise the animal's resistance to infection [19]. Thus, the aim of the probiotic approach is to repair the deficiencies in the microflora and restore the animal's resistance to disease. Such a treatment does not introduce any foreign chemicals into the animal's internal environment and does not run the risk of contaminating the carcass and introducing hazardous chemicals into the food chain. Probiotics are now replacing the chemical growth promoters for farm animals and claims have also been made for increasing resistance to disease. Probiotics are preparations based on live microorganisms that are consumed as food and feed additives, and which have a beneficial effect on the health status of humans or animals. It is said that probiotics help prevent imbalances, and enhance the growth of the healthy microflora [17]. Besides, probiotics are widely produced, promoted and marketed.

Microorganisms used as probiotics in animal nutrition: Most probiotic products utilize one or more of several types of bacteria. The most commonly used bacterial probiotics are the strains of Bifidobacterium, Enterococcus, Lactobacillus, Bacillus, Pediococcus and Streptococcus. Some products contain viable yeast and other fungi in addition to bacteria. There are marked differences between the various probiotic groups regarding their properties, origin and mode of action.

The following table shows probiotics that are commonly used in animal nutrition summarized from many l iteratures [2,9,15,17,20,24].

The use of probiotics as farm animal feed supplements dates back to the 1970's. They were originally incorporated into feed to increase the animal's growth and to improve its health by increasing its resistance to disease [19]. Yet, use of probiotics in humans and animal species such as young pigs has been widely reported in the scientific literature. Numerous studies have shown that humans or animals fed probiotics have altered intestinal bacterial populations, improved resistance to disease, reduced shedding of pathogens when challenged orally, increased intestinal immunity, reduced disease symptoms, and improved health [15].

Certain microorganisms which are intentionally added to the feed (probiotics) counteract possible disruptions of the equilibrium and lead to eubiosis. Thus the colonisation of the intestine by undesirable microorganisms can be suppressed. As yet, not all actions of probiotics have been satisfactorily explained by science. Their overall positive effects, based on developing metabolic activity, comprise both direct and especially indirect effects (Figure 2) [9].

Probiotic is a generic term, and products can contain yeast cells, bacterial cultures, or both that stimulate microorganisms capable of modifying the gastrointestinal environment to increase health status and improve feed efficiency.

Administration of probiotic strains separately and in combination significantly improved feed intake, FCE, daily weight gain and total body weight in chicken, pig, sheep, goat, cattle and equine.

**Table 1.** Probiotics commonly used in animal feed.

| Genus | Species |
|---|---|
| *Bifidobacterium* | *B. lactis*<br>*B. longum*<br>*B. pseudolongum*<br>*B. thermophilum* |
| | *B. bifidum* |
| *Enterococcus* | *E. faecalis* |
| *Lactobacillus* | *L. acidophilus*<br>*L. amylovorus*<br>*L. brevis*<br>*L. casei*<br>*L. farmicinis*<br>*L. fermentum*<br>*L. plantarum*<br>*L. reuteri*<br>*L. rhamnosus* |
| *Lactococcus* | *L. lactis* |

| | |
|---|---|
| *Leuconosto* | *L. citreum*<br>*L. lactis*<br>*L. mesenteroides* |
| *Pediococcus* | *P. acidilactici*<br>*P. pentosaceus subsp. Pentosaceous* |
| *Streptococcus* | *S. infantarius*<br>*S. salivarius* |
| *Bacillus* | *B. cereus*<br>*B. licheniformi*<br>*B. subtilis* |
| *Saccharomyces* | *S. cerevisiae (S. boulardii)*<br>*S. pastorianus (S. carlsbergensis)* |
| *Aspergyllus* | *A. oriza*<br>*niger* |

In monogastric animals, strains of Lactobacilli, Bacillus subtilis and Streptococci have been used as probiotics. The use of these probiotics has been shown to help improve gain and feed efficiency in poultry and swine. However, other researchers have observed no significant response in swine [13].

In ruminant animals, the application of yeast (Saccharomyces cerevisiae) in the form of live culture, or dead cells with culture extracts, has proved successful in beneficially modifying rumen fermentation [5]. Probiotics have been used to potentially replace or decrease the use of antibiotics in neonatal and stressed calves, to enhance milk production in dairy cows, and to improve daily gain and feed efficiency in beef cattle [26]. However, performance results of experiments where cattle were fed probiotics are generally inconsistent.

# 4. EFFECTS OF PROBIOTICS ON ANIMAL HEALTH AND PERFORMANCE

## 4.1 Effects of probiotics on cattle health and performance

Though study of probiotics for human and animal use has increased since the mid 1950's, Yoon and Stern [27] explained that the study of production responses by growing and lactating ruminants, and interest in the corresponding mode of action of probiotics, is more recent. Indeed, the increased interest about pathogen contamination of meat and meat products has resulted in a recent surge of experiments evaluating the efficacy of probiotics [1].

In ruminants, yeast cultures can stimulate forage intake by increasing the rate of digestion of fibre in the rumen in the first 24 h after its consumption. This improvement in early digestion and intake is brought about by alterations in the numbers and species of microorganisms in the rumen. The increase in forage intake can result in improved live weight gain, milk yield and milk fat content, although the effects are often small in dairy cows [3].

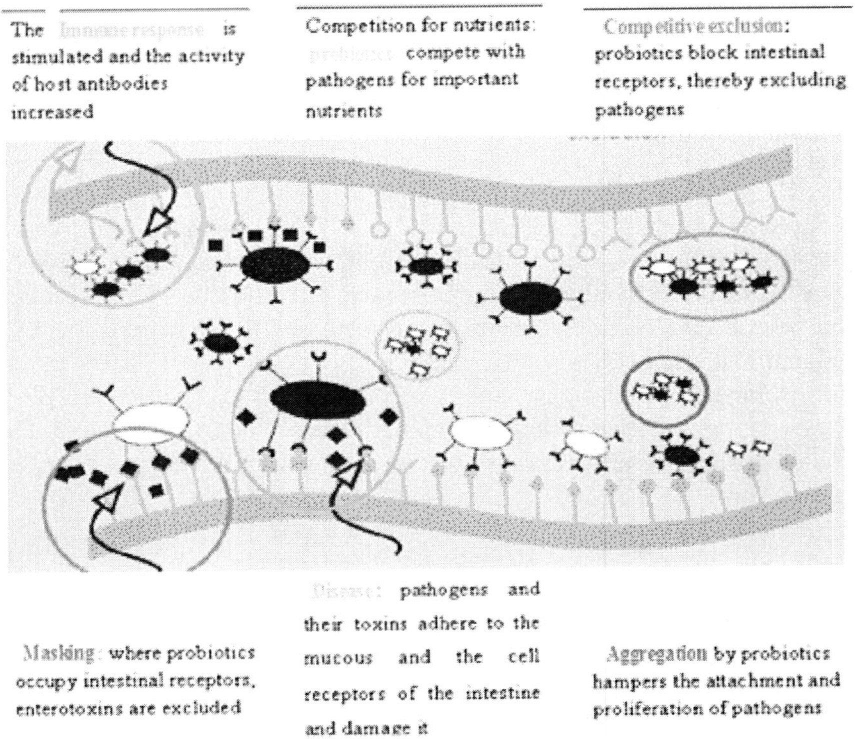


**Figure 2.** Microbiological interactions in the intestine [9].

## 4.2 Pre-ruminant calves

Though there are published studies that have evaluated various DFM formulas in the diet of un-weaned calves, the effect of DFM on young calves is much less clear [15].

Based on studies in the newborn dairy calf, Newman and Jacques [28] concluded that the normal population of gastrointestinal microbes is extremely beneficial for preventing infection. In support, Fuller [29] observed that germ-free animals are much more susceptible to disease than are inoculated animals. Different results suggest that probiotics are beneficial for establishing and maintaining a positive microbial balance in newborn calves by decreasing the prevalence of coliforms. Moreover, a decreased incidence of diarrhea has generally been associated with an increased shedding of Lactobacillus [30]. In the contrary, the Lactobacillus population decreases when animals are stressed, and stress often leads to an increased incidence of diarrhea in neonates [1]. In fact, there are results that report feeding probiotics to dairy calves had no effect on fecal scores or oocyst shedding [31].

During the first 3 weeks of preruminant's life, decreasing the incidence of diarrhea is most likely a more important response than performance responses, as enteric disease is most prevalent during this period [26].

Reports about the response of feeding probiotics in performance are highly variable. There are many studies that reported no improvement in daily gain as a result of feeding DFM to dairy calves [30]. In contrast, others reported increased growth and feed efficiency by calves treated with probiotic bacteria [32,33].

Hence, these benefits were hypothesized to result from improvement of intestinal conditions because of lower fecal scores (less scouring) in calves fed probiotics. Furthermore, Adams et al. [33] suggested that the increase in body-weight gain is not only during the milk feeding period but also after weaning. Finally, Quigley [15] suggested that, variation in responses is likely a function of interactions between diet, the degree of the pathogen challenge and other stressors.

## 4.3 Receiving cattle

Weaning, transport, fasting, assembly, vaccination, castration and dehorning are some of the stresses beef calves undergo upon entering the feedlot. These stresses result in decreased performance and increased morbidity and mortality, in part due to altered microorganisms in the rumen and lower gut. Therefore, feeding probiotics might reduce these changes in the microbial population. Krehbiel et al. [25] summarized data from several research trials, and concluded that feeding a probiotics at processing, throughout the receiving period (average of 30 days), or both resulted in a 13.2% increase in daily gain, 2.5% increase in feed consumption, and a 6.3% improvement in feed: gain. The greatest performance response to the probiotics generally occurred within the first 14 days of the receiving period. As well, morbidity was decreased by 27.7% in cattle receiving the probiotics compared with control cattle.

**Table 2.** Main Targets for probiotics' use in different types of animals [25].

| Animals | Common Benefits |
|---|---|
| Pig | Improve colostrum quality, milk quality and quantity<br>Increase litter size and vitality<br>Increase piglet weight<br>Reduce risk of diarrhea<br>Improve feed efficiency, diet digestibility and meat quality<br>Limit constipation<br>Decrease stress |
| Poultry | Increase body weight gain<br>Reduce mortality<br>Increase carcass quality decreasing contamination<br>Increase bone quality |
| Veal Calf | Promote weight gain and optimal maturation of rumen microbiota limiting acidosis<br>Increase feed efficiency, milk yield, quality and digestive safety at weaning<br>Reduce risk of pathogen colonization and limit shedding of human pathogens |
| Horse | Improve diet digestibility, milk quality and quantity<br>Limit diarrhea |
| | Avoid hindgut disorders (acidosis, colic)<br>Limit stress (Transportation, race etc.) |

**Table 3.** Influence of various probiotics on the performance of animals [9].

| Production branch | DWG (% of control) | FCR (% of control) |
|---|---|---|
| Piglet production | +4.8 | -1.5 |
| Calf production | +5.4 | -2.5 |
| Growing/fattening pigs | +3.7 | -5.1 |
| Growing/fattening cattle | +3.4 | -2.7 |

## 4.4 Feedlot cattle

The addition of yeast to intensive beef diets has increased daily live weight gain and FCE [3].

Growth: In relation with the growth of cattle, Swinney-Floyd et al. [34] reported that though bacterial DFM did not alter DM intake, ADG for the animals inoculated with the combination of Propionibacterium and L. acidophilus was greater than control calves. In line with this, Huck et al. [38] reported a tendency for greater carcass-adjusted ADG in heifers that were sequentially fed a Lactobacillus and then a Propionibacterium or a Propionibacterium and then a Lactobacillus probiotics between the receiving and finishing phases compared with control heifers. But, in contrast, Rust et al. [35] reported no difference for carcass-adjusted ADG in control steers versus steers supplemented with Lactobacillus and Propionibacterium. However, ADG for the average of all probiotics - treated steers was increased 6.2% over control steers. In addition, feed efficiency for the entire feeding period was improved for steers receiving certain probiotics treatments compared with control steers. Galyean et al. [36] also reported that the final BW, ADG and hot carcass weight were significantly greater for steers treated with probiotics versus the control animals. Similarly, McPeake et al. [37] reported steers fed a probiotics had greater final weight, ADG, DM intake, hot carcass weight and carcass-adjusted ADG compared with control steers. In addition, there was also a trend for improved feed efficiency as the concentration of supplemental L. acidophilus increased.

Carcass: With regards to carcass characteristics, different studies [35,37] indicated that, the addition of probiotics did not affect yield grade, quality grade, dressing percentage, marbling score. However, hot carcass weight was generally greater when probiotics were fed. On the other hand, Huck et al. [38] reported that the percentage of carcasses graded as was greater in heifers receiving a Propionibacterium probiotics in both the receiving and finishing phases than in control heifers. To this end, it can be seen that improvements in carcass characteristics as a result of probiotics are questionable, except for hot carcass weight. Because probiotics generally improve ADG, hot carcass weight would be increased if probiotics supplementation increased ADG and final BW [39]. Beauchemin et al. [1] elaborated that the positive effects of probiotics on performance might be associated with a decrease in ruminal acidosis and/or improved microbial balance in the lower gastrointestinal tract.

Milk yield and composition in dairy cows: Researches concerning the effects of probiotics in dairy cows are limited. Even in the available studies the probiotics were fed together with other additives making it difficult to judge the effects of the probiotics themselves. In the few studies available in which probiotics were fed to dairy cows, the milk yield increased by 0.75–2.0 kg/d. In general, an increased milk yield has been a consistent response, whereas changes in milk composition have been variable. In support with this, Gomez-Basauri et al. [40] described that cows fed lactic acid bacteria and mannan-oligosaccharide produced more milk. The authors reported that milk yields increased over time for lactic acid bacteria-and mannanoligosaccharide-fed cows, whereas control cows maintained constant milk yields. Furthermore, there are experiments that suggest probiotics fed alone or in combination with fungal cultures might be efficacious for increasing milk production by lactating dairy cows [41,42]. However, more research is needed before recommendations can be made have been conducted with combinations of fungal cultures and lactic acid bacteria.

Effects of probiotics on poultry health and performance: Addition of probiotics has shown beneficial effects on growth performance of poultry. In broilers, supplementation of a diet with probiotics has resulted in improved feed conversion rate and average live weight in comparison to the control group. Moreover, in a study by Mountzouris et al. administration of the multi-strain probiotic in the drinking water significantly increased average daily weight gain and feed efficiency and numerically reduced mortality rate in comparison with a negative control.

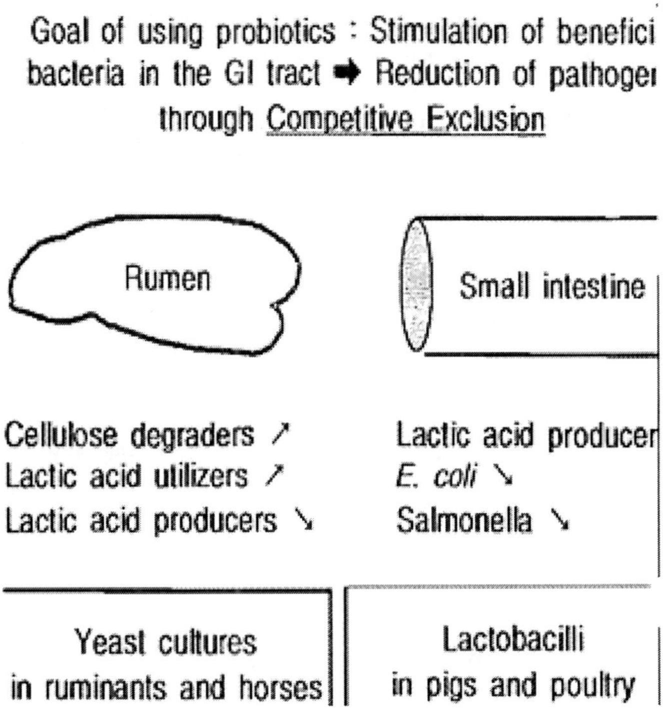


**Figure 3.** How probiosis can have a beneficial effect on animal nutrition.

**Table 4.** Performance values of experimental broiler chick for control and probiotic dietary treatments at the age of 28-42 days.

| Performance Parameters | Control | Probiotic |
|---|---|---|
| Feed intake (g/d) | 1008.42 | 1119.72 |
| FCR (g feed/g meat) | 2.03 | 1.82 |
| Body weight gain (g) | 547.13c | 920.96a |
| Mortality rate % | 15.0b | 2.5a |

Effects of probiotics on pig health and performance: Some reports have indicated that supplementation of probiotics improves performance in suckling pigs, weanling pigs, grower pigs and finishing pigs. Administration of the multi-strain probiotics tended to cause higher weight gain and feed efficiency. In support with this, Giang et al. reported that piglets fed probiotic complexes diets had higher feed intake, daily gain and better feed conversion during the 1st 2 weeks after weaning. This indicates a synergistic effect of different probiotic strains under in vivo conditions. Eventually, as a result of feeding a fermented diet, the time available for the gastrointestinal microflora to decarboxylate free amino acids present in the diet minimizes which has shown to improve performance in pigs [43].

In addition to the improved average daily gain, addition of probiotics reduced mortality rates of growing and finishing pigs. In line with this, different studies revealed that the incidence and severity of diarrhea as well as mortality rate were significantly decreased as a result of feeding probiotics. Moreover, live yeast supplementation to the diet of pigs has resulted in demonstrable reductions in the quantity of pathogenic bacteria.

Application of probiotics and recommendations for use: Although some products contain purified strains of individual organisms, most probiotics products are a combination of several species of bacteria and yeast and other fungi. Most probiotics products for calves are sold as feed additives, which are added to milk or milk replacer just prior to feeding, while others are administered as gels, pastes, or boluses. But, their application on the farm is more challenging. Adding probiotics to pelleted feeds is difficult, as temperatures and pressures used in pelleting generally kill most organisms [15].

The stabilisation of the digestive or microflora in ruminants and in monogastric animals can only be effectively achieved by continuous supplementation of the feed with probiotics because the microorganisms used in animal nutrition do not permanently colonise the intestine. Increased short-term supplementation of probiotics may be useful under certain conditions but should be followed by continuous supplementation thereafter. General guidelines on the optimal dosage and the period of supplementation are not possible because factors such as stability of the probiotic in the feed and in the digestive tract, the specific mode of action of the microbes contained in the product and the status of the intestinal microflora in the host all modulate the effect of the corresponding product [9].

The optimal dose must be determined individually for each product and each target species in feeding trials. The rate of inclusion given by the manufacturer, therefore, is based on information gained from efficacy studies [9]. In general, however, it is accepted that the inclusion rate of all probiotics should be higher when the intestinal microflora is unstable and particularly when for ruminant the diet composition contain high rapid-fermentable sugar which can entail sub-acidosis. In addition, the overall consumption of probiotics by older animals will be higher because of a higher feed intake compared to younger animals. Therefore, with continuous supplementation, the inclusion rate may be reduced during the growth of the animals without the concentration of the probiotic microorganisms in the intestine dropping below the level of efficacy.

Efficacy of probiotics: Reports of probiotics efficacy are variable but this may depend greatly on survival rate of strains, varying stabilities of strains, low probiotic doses, frequency/infrequency of administration, interactions with some medicines, health and nutritional status of the animal and the effect of age, stress, genetics and type differences of animals [14]. To be effective, the desirable microorganism should not be harmful to the host animal, should be resistant to bile and acid, should colonise the gut efficiently, should inhibit pathogenic activity, and should be viable and stable under manufacturing and storage conditions [3].

Research points to the fact that probiotics are most effective in animals during microflora development or when microflora stability is impaired. The benefit of probiotics with respect to health status and performance is expected to be highest in young animals such as piglets, newly-hatched chickens or calves, because these animals have not yet developed a stable gut microflora. Moreover, when animals undergo therapeutic treatment of diseases with antibiotics, the gut microflora is generally decimated. Therefore, administration of probiotics after antibiotic treatment assists in re-establishing a beneficial gut microflora to prevent the host from recurrent pathogenic colonisation [20].

For pigs, it is suggested that the effects of probiotics appear to be more consistent and positive in piglets rather than in growing finishing pigs. Hence, in a review of the response of pigs of various ages to the administration of probiotics, it was concluded that probiotics were effective for young pigs, in which the digestive tract is still developing after weaning. However, probiotics were less effective for growing and finishing pigs, which already have a balanced population of microorganisms [3].

With regard to cattle, similarly, improvements in animal performance may be limited in young, milk-fed calves. Rather, it appears that probiotics may be most useful under specific conditions whereby calves are exposed to immune or management challenges that may disrupt the intestinal environment. Under stress conditions, probiotics may reduce the risk of scours caused by an upset in the normal intestinal flora of calves [5]. As in the neonatal calf, the response to probiotics might be greater if administered to newly weaned and/or received beef calves, which are more prone to health problems. Gill et al. [44] suggested that extremely healthy calves and extremely sick calves might be less likely to respond to probiotics treatment.

## 4.5 Consequences of using probiotics growth promoters

Safety for humans: Comprehensive studies have shown that there are no hazards for probiotic users. Direct contact of registered probiotic products with skin, mouth and nose do not compromise human health. In model trials it has been established that even long-term or increased exposure do not constitute a risk to health. As a food consumer, however, man does not come into contact with the probiotics fed to the animal. As probiotics are administered exclusively via the feed, their action is restricted to the gastro-intestinal tract. Hence they are not absorbed, they cannot be transferred into foodstuffs of animal origin and hence do not lead to residues [9].

Safety for the environment: Having exerted their effect in the digestive tract, the probiotic reaches the exit of the intestine in the digesta, together with other intestinal microorganisms. On their way along the digestive tract the majority of the probiotic bacteria die off, since their growth and proliferation is severely restricted by competition from other microorganisms present in the large intestine [9]. The probiotics are already partly broken down and digested like other organic nutrients in the intestine so that only a small proportion is excreted viable in the faeces and survives in the manure to reach fields and grassland. Evidence of the harmlessness of the probiotic to the environment is one important subject for its registration. In general, any negative impact is highly unlikely since all these microorganisms are derived from nature.

Safety for animals: Overall, the microorganisms approved for animal nutrition have a very good safety record. Probiotics do not constitute any health hazard for the animal. Since they are not transferred from the intestine into the body of the animal, they do not affect any metabolic processes, nor do they have any negative impact on the animal [9].

Contrary to this, Hughes and Heritage [7] questioned the usefulness of probiotics as it remains unproven. Though, probiotics do have some strong supporters in the scientific community, these are matched by an equal number of detractors. As far as the beneficial effects of probiotics, they have been demonstrated almost exclusively under defined experimental conditions. In addition, there is the lack of evidence as to the mechanism of action and of the effects on host animals.

## REFERENCES

1. Beauchemin KA, Krehbiel CR, Newbold CJ (2006) Enzymes, Bacterial Direct-Fed Microbials and Yeast: Principles for Use in Ruminant Nutrition. In: Mosenthin R, Zentek J, Ebrowska TZ (Eds) Biology of Nutrition in Growing Animals.

2. Fanelli A (2012) Direct-Fed Microbials (DFMs) in horses and poultry: effects on digestibility, nutritional value of animal products and animal health. Graduate School of Veterinary Sciences for Animal Health and Food Safety. Doctoral Program in Animal Nutrition and Food Safety.Università degli Studi di Milano.

3. McDonald P, Edwards RA, Greenhalgh JFD, Morgan CA, Sinclair LA et al (2010), Animal Nutrition (7 th edn.) Pearson Books.

4. Chahal US, Niranjan PS, Kumar S (2008) Handbook of General Animal Nutrition. International Book Distributing Co. India, pp2 292.

5. Cho JH, Zhao PY, Kim IH (2011) Probiotics as a Dietary Additive for Pigs. Journal of Animal and Veterinary Advances 10 pp2 2127-2134.

6. Simon O (2005) Micro-Organisms as Feed Additives–Probiotics. Advances in Pork Production 16: p 2 161.

7. Hughes P, Heritage J (2002) Food and Agriculture Organization. Antibiotic growth-promoters in food animals. Retrieved from Leeds, U.K. website: http://www.fao.org/DOCREP/ARTICLE/AGRIPPA/555_EN.HTM

8. Piva A, Galvano F, Biagi G, Casadei G (2006) Intestinal Fermentation: Dietary and Microbial Interactions. In: Mosenthin R, Zentek J, Ebrowska TZ (Eds). Biology of Nutrition in Growing Animals, p 1-31.

9. FEFANA (2005) Probiotics in Animal Nutrition. EU feed additives and Premixtures Association.

10. Roberfroid MB (1998) "Prebiotics and synbiotics: concepts and nutritional properties."The British journal of nutrition 80: 197-202.

11. Tannock GW (1999) Probiotics: A Critical Review. Horizon Scientific Press, Wymondham.

12. Luchansky JB (2000) Use of biotherapeutics to enhance animal well being and food safety. In: Piva G, Masoero F (Eds). Feed Production Conference, Piacenza, Italy, pp. 188–194.

13. Pollmann DS, Danielson DM, Peo ER (1980) Effects of Microbial Feed Additives on Performance of Starter and Growing-Finishing Pigs 1. Journal of Animal Science 51: 577-581.

14. Lawrence TLJ and Fowler VR (2002) ‘Growth Promoters’, Performance Enhancers, Feed Additives and Alternative Approaches. Pp2 320-329.

15. Quigley J (2011) Direct-Fed Microbials (Probiotics) in Calf Diets. A BAMN Publication.

16. Chaucheyras F and Durand H (2010) Probiotics in animal nutrition and health. Beneficial Microbes 1: 3-9.

17. Jiménez G (2012) Probiotics in Animal Nutrition - a Century of Research.

18. Guarner F, Malagelada JR (2003) Gut flora in health and disease, Lancet, 512–519pp.

19. Fuller R (1992) "History and Development of Probiotics." Springer Netherlands, 1-8.

20. Steiner T (2009) Probiotics in Poultry and Pig Nutrition: Basics and Benefits. The Poultry site.

21. Umesaki Y, Okada Y, Imaoka A, Setoyama H and S. Matsumoto (1997) Interactions between Epithelial Cells and Bacteria, Normal and Pathogenic. Science Magazine, 9.

22. Yu H, Braun P, Yildirim MA, Lemmens I, Venkatesan K et al. (2008) High-Quality Binary Protein Interaction Map of the Yeast Interactome Network. Science 322: 104-110.

23. Alakomi HL, Saarela M, Helander IM (2003) Effect of EDTA on Salmonella enterica serovar TyphimuriumInvolves a Component not Assignable to Lipopolysaccharide Release. Microbiology 149: 2015-2021.

24. EPA (2012) Probiotics in Animal Nutrition, Today and Tomorrow: The Experts' Opinion.

25. Ahasan ASML, Agazzi A, Invernizzi G, Bontempo V, Savoini G (2015) The Beneficial Role of Probiotics in Monogastric Animal Nutrition and Health. Journal of Dairy, Veterinary & Animal Research 2.

26. Krehbiel CR, Rust SR, Zhang G, Gilliland SE (2003) Bacterial direct-fed microbials in ruminant diets: Performance response and mode of action. J. Anim. Sci 81.

27. Yoon IK, Stern MD (1995) Influence of direct-fed microbials on ruminal microbial fermentation and performance of ruminants: A review. Asian-Austr. J. Anim. Sci 8: 533–555.

28. Newman KE, Jacques KA (1995) Microbial feed additives for pre-ruminants. In: Wallace RJ, Chesson A (Eds), Biotechnology in Animal Feeds and Animal Feeding. VCH, Weinheim, Federal Republic of Germany, pp2. 247–258.

29. Fuller R (1989) A review: Probiotics in man and animals. J. Appl. Bacteriol. 66: 365–378.

30. Ellinger DK, Muller LD, Glantz PJ (1980) Influence of feeding fermented colostrum and Lactobacillus acidophilus on fecal flora of dairy calves. J Dairy Sci 63: 478-482.

31. Harp JA, Jardon P, Atwill ME, Zylstra R, Checel S et al. (1996) Field testing of prophylactic measures against Cryptosporidium parvum infection in calves in a California dairy herd. Am. J. Vet. Res 57: 1586–1588.

32. Abe F, Ishibashi N, Shimamura S (1995) Effect of Administration of Bifidobacteria and Lactic acid bacteria to newborn calves and piglets. J. Dairy Sci 78: 2838–2846.

33. Adams MC, Luo J, Rayward D, King S, Gibson R et al. (2008) Selection of a Novel Direct-fed Microbial to Enhance Weight Gain in Intensively Reared Calves. Anim. Feed Sci. Technol 145:41-52.

34. Swinney-Floyd D, Gardner BA, Owens FN, Rehberger T, Parrott T (1999) Effect of inoculation with either strain P-63 alone or in combination with Lactobacillus acidophilus LA53545 on performance of feedlot cattle. J. Anim. Sci. 77.

35. Rust SR, Metz K, Ware DR (2000) Effects of Bovamine TM rumen culture on the performance and carcass characteristics of feedlot steers. Mich. Agric. Exp. Sta. Beef Cattle, Sheep and Forage Syst. Res. Dem. Rep. 569: 22–26.

36. Galyean ML, Nunnery GA, Defoor PJ, Salyer GB, Parsons CH (2000) Effects of live cultures of Lactobacillus acidophilus (Strains 45 and 51) and Propionibacterium freudenreichii PF-24 on performance and carcass characteristics of finishing beef steers.

37. McPeake CA, Abney CS, Kizilkaya K, Galyean ML, Trenkle AH et al. (2002) Effects of direct-fed microbial products on feedlot performance and carcass characteristics of feedlot steers, pp2. 133.

38. Huck GL, Kreikemeier KK, Ducharme GA (2000) Effect of feeding two microbial additivesin sequence on growth performance and carcass characteristics of finishing heifers.

39. Elam NA, GleghornJF, Rivera JD, Galyean ML, Defoor PJ, et al. (2003) Effects of live cultures of Lactobacillus acidophilus (strains NP45 and NP51) and Propionibacterium freudenreichii on performance, carcass, and intestinal characteristics, and Escherichia coli strain O157 shedding of finishing beef steers. J. Anim. Sci 81: 2686–2698.

40. Gomez-Basauri MBJ, de Ondarza, Siciliano-Jones (2001) Intake and milk production of dairy cows fed lactic acid bacteria and mannanoligosaccharide. J. Dairy Sci 84: 283.

41. Komari RK, Reddy YKL, Suresh J, Raj DN (1999) Effect of feeding yeast culture (Saccharomyces cerevisiae) and Lactobacillus

acidophilus on production performance of crossbred dairy cows. J. Dairy Sci. 82: 128.

42. Block E, Nocek JE, Kautz WP, Leedle JAZ (2000) Direct Fed Microbial and Anionic Salt Supplementation to Dairy Cows Fed 21 Days Pre- to 70 Days Postpartum. J. Anim. Sci. 78: 304.

43. Pedersen LG, Castelruiz Y, Jacobsen,, Aasted B (2002) Identification of monoclonal antibodies that cross-react with cytokines from different animal species. Vet Immunol Immunopathol. 88: 3-4.

44. Gill DR, Smith RA, Ball RL (1987) The effect of probiotic feeding on health and performance of newly-arrived stocker calves. Okla. Agr. Exp. Sta. MP-119, 202–204.

# CHAPTER 2

# Comparative performance of broiler chickens offered ten equivalent diets based on three grain sorghum varieties as determined by response surface mixture design

**S.Y. Liu[a], H.H. Truong[a, b], A. Khoddami[c], A.F. Moss[a], P.C. Thomson[d], T.H. Roberts[c], P.H Selle[a]**

[a] Poultry Research Foundation, Faculty of Veterinary Science, The University of Sydney, 425 Werombi Road, Camden, NSW 2570,Australia

[b] Poultry Cooperative Research Centre, PO Box U242, University of New England, Armidale, NSW 2351, Australia

[c] Plant Breeding Institute, Faculty of Agriculture and Environment, University of Sydney, NSW 2006, Australia

[d] Faculty of Veterinary Science, The University of Sydney, 425 Werombi Road, Camden, NSW 2570, Australiaa

## HIGHLIGHTS

- Growth performance of chickens was determined by response surface design.
- Kafirin, phenolic compounds and phytate depressed energy utilisation.
- Free ferulic acid influenced distal ileal protein disappearance rate.

- Low protein sorghum may advantage growth performance of broiler chickens.
- Response surface mixture design predicted the best sorghum combinations.

## ABSTRACT

Ten nutritionally-equivalent diets containing 620 g/kg grain sorghum, based on seven blends and three varieties (Block I, HP, Liberty), were offered to a total of 360 male Ross 308 chicks from 7 to 28 days post-hatch. The effects of dietary treatments on growth performance, nutrient utilisation, starch and protein (Nitrogen) digestibility coefficients, disappearance rates and starch:protein disappearance rate ratios in the distal jejunum and distal ileum were determined. On the basis of response surface contour plots, the optimal weight gain of 1439 g/bird would be generated by a 79.2-20.8 blend of HP and Liberty and the optimal feed conversion ratio (FCR) of 1.486 by a 55.6-44.4 HP-Liberty blend. Similarly, the optimal nitrogen-corrected apparent metabolisable energy (AMEn) of 12.38 MJ would be generated by a 32.0-68.0 blend of HP and Liberty and the optimal apparent metabolisable energy to gross energy (AME:GE) ratio of 0.845 by a 31.1-68.9 HP-Liberty blend. Thus Block I sorghum was effectively rejected. Instructively, Block I contained higher concentrations of kafirin, phenolic compounds and phytate than the two other sorghum varieties. Across the ten diets, sorghum concentrations of these three factors were negatively correlated with AME:GE ratios or efficiency of energy utilisation. These negative linear regressions indicate that all three factors may be contributing to sub-standard starch/energy utilisation in broiler chicks offered sorghum-based diets.

## KEYWORDS

Broiler; Kafirin; Nutrition geometry; Response surface design; Sorghum

# ABBREVIATIONS

AME - apparent metabolisable energy

AMEn - nitrogen-corrected apparent metabolisable energy

AIA - acid insoluble ash

FCR - feed conversion ratio

N - nitrogen

NC - negative control

P - phosphorus

# 1. INTRODUCTION

At times, sorghum provides, either partially or entirely, the feed grain base in Australian pig and poultry diets but it has been associated with sub-optimal or inconsistent growth performance in broiler chickens (Selle et al., 2010a and Liu et al., 2015a). A beneficial characteristic for including sorghum in broiler diets is as a source of energy, which is mainly derived from starch. However, energy utilisation in sorghum-based diets is considered to be inferior to wheat and maize under Australian conditions (Black et al., 2005). Liu et al. (2015a) nominated three factors inherent in sorghum which may be largely responsible for the relatively inefficient utilisation of dietary energy. The three factors were (i) kafirin, the dominant protein fraction in grain sorghum; (ii) phenolic compounds, potentially including condensed tannin and (iii) phytate or *myo*-inositol hexaphosphate (IP6). In order to clarify the underlying reasons for the inconsistent growth performance and nutrient utilisation in broiler chickens offered sorghum-based diets, the present study embraces 'equilateral triangle response surface design' (Raubenheimer, 2011 and Liu et al., 2015b). Such designs generate response surfaces which graphically depict interactions between feedstuffs, in this case grain sorghum. The total amount of the mixture must be held constant so that responses are entirely dependent on the relative proportions of feedstuffs in the blend. In the present study, three diverse sorghum varieties were extensively characterised and incorporated into ten nutritionally-equivalent diets at 620 g/kg. The diets contained different concentrations of kafirin, phytate and phenolic compounds by mixing the three sorghums at different inclusion rates. 'Contour plot' response surfaces were then used to interpret variations in broiler performance. The intention of the present study was to examine how three different sorghums, with divergent nutritional characteristics, influence growth performance and nutrient utilisation by using equilateral triangle mixture design.

# 2. MATERIALS AND METHODS

## 2.1. Characteristic of sorghums

Two red (HP and Block I) and one white (Liberty) grain sorghum varieties were evaluated in this study. "Block I" was grown in the Murrumbidgee Irrigation Area (NSW, Australia) in 2012, "HP" on the Liverpool Plains (NSW, Australia) in 2013 and the white variety, Liberty, was grown on the Darling Downs (Qld. Australia) and harvested in 2014. These three sorghums were extensively characterised which included concentrations of protein, amino acids, kafirin, minerals, phytate, NIR AusScan results and grain texture (Symes, 1965) as shown in Table 1. Concentrations of protein and fourteen amino acids except cysteine and tryptophan, were determined in duplicates as outlined by Cohen and Michaud (1993). Kafirin was quantified by a method adapted from Wallace et al. (1990)and Hamaker et al. (1995). Ground sorghum (100 mg) was incubated with 1 mL total protein extraction buffer (12.5 mM sodium borate, pH 10.0, 1% SDS, 2% 2-mercaptoethanol) for 1 h with agitation at room temperature. Each sample was centrifuged for 15 min at 14,000 rpm at room temperature. Total protein was extracted twice more as above and the supernatants pooled. To isolate kafirin from the total protein extract, 100% t-butanol was added to the pooled supernatant at a final concentration of 60%, and incubated for 30 min at room temperature with occasional mixing. The extract was then centrifuged for 15 min at 9500 rpm at room temperature, which resulted in the formation of a pellet (non-kafirin) and supernatant (alcohol-soluble kafirin). Kafirin was extracted from each sorghum sample in duplicate.

The concentrations of nine minerals in the diets and excreta were analysed by Inductively Coupled Plasma-Optical Emission Spectroscopy using a Perkin Elmer OPTIMA 7300 (Perkin Elmer Inc., Waltham, MA).

Phytate was determined by an HPLC procedure (Skoglund et al., 1997a and Skoglund et al., 1997b). Additionally, as per Table 2, concentrations of total phenolic compounds, polyphenols, free, conjugated, and bound ferulic acid and total phenolic acids plus RVA starch pasting profiles in the three sorghums were determined. The rapid visco-analysis (RVA) starch pasting profiles were determined using a Rapid-Visco-Analyzer. Within 13 min intervals, a 28 g mixture of sorghum or diet and water (15:85 w/w) were prepared and held at 50 °C temperature for 1 min and then heated from 50 °C to 95 °C. After holding the hot paste at 95 °C for 2.5 min, the slurry was again cooled to 50 °C, and then held at that temperature for 2 min. The analytical procedures for quantification of phenolic compounds were previously outlined in detail inKhoddami et al. (2015).

## 2.1.1. Diet preparation

Three diets containing 620 g/kg Block I, HP or Liberty varieties were formulated with an energy density of 12.94 MJ/kg and similar amino acid profiles. Their dietary composition and calculated nutrient specifications of the three basal diets are tabulated in Table 3. The equilateral triangle trial design for ten broiler diets based on blends of three basal diets are shown in Fig. 1. The sorghum grains were ground through a 3.2 mm hammer-mill screen prior to their dietary incorporation. Diet 1, 2 and 3 were mixed individually and then remixed to form the remaining experimental diets according to the percentages as described in Fig. 1. The diets were pelleted through a Palmer PP330 pellet press (Palmer Milling Engineering, Griffith, NSW, Australia) at a conditioning temperature of 84 °C by the automatically controlled introduction of steam into the conditioner with a residence time of 14 s. The conditioning temperature was recorded by thermal probes at the exit of the conditioner before the diet entered the pellet press with die dimensions of 4 mm in diameter. Finally, the diets were cooled in a vertical cooler to room temperature and then crumbled.

**Table 1.** NIR AusScan results, concentrations (as-is) of protein, amino acids, kafirin, minerals and phytate, grain texture of Block I, HP and Liberty sorghum varieties.

| Item | Block I | HP | Liberty |
|---|---|---|---|
| NIR AusScan (g/kg) | | | |
| AME (MJ/kg) | 13.9 | 14.2 | 14.3 |
| NIR Protein | 142 | 113 | 82 |
| Total starch | 756 | 818 | 851 |
| Crude fibre | 37 | 30 | 28 |
| Acid detergent fibre | 102 | 97 | 85 |
| Neutral detergent fibre | 140 | 104 | 135 |
| Total soluble NSP | <0.1 | <0.1 | <0.1 |
| Total insoluble NSP | 30 | 57 | 68 |
| Insoluble arabinoxylans | 36 | 43 | 57 |
| Hydration capacity (%) | 9.9 | 8.6 | 9.2 |
| Analysed Amino acid profile (g/kg) | | | |
| Crude protein | 137.1 | 109.1 | 80.9 |
| Arginine | 4.5 | 4.2 | 2.8 |
| Histidine | 2.9 | 2.4 | 1.8 |
| Isoleucine | 5.1 | 4.2 | 3.6 |
| Leucine | 17.7 | 14.2 | 10.4 |
| Lysine | 2.7 | 2.5 | 1.9 |
| Methionine | 2.0 | 1.6 | 1.1 |
| Phenylalanine | 6.8 | 5.6 | 4.2 |
| Threonine | 4.2 | 3.5 | 2.7 |
| Valine | 6.6 | 5.6 | 4.1 |
| Alanine | 12.0 | 9.5 | 7.1 |
| Aspartic acid | 8.5 | 7.2 | 5.4 |
| Glutamic acid | 28.3 | 23.2 | 16.8 |
| Glycine | 4.0 | 3.5 | 2.7 |
| Proline | 11.2 | 8.8 | 6.6 |
| Serine | 6.1 | 4.9 | 3.8 |
| Tyrosine | 2.3 | 3.2 | 1.5 |

| | | | |
|---|---|---|---|
| Kafirin | 67.1 | 50.5 | 41.4 |
| Minerals (mg/kg) | | | |
| Calcium | 186 | 156 | 146 |
| Phosphorus | 3644 | 3122 | 1941 |
| Copper | 5.76 | 2.15 | 1.55 |
| Iron | 46.3 | 34.7 | 64.6 |
| Potassium | 3748 | 3219 | 2482 |
| Magnesium | 1710 | 1516 | 1047 |
| Manganese | 284 | 18.8 | 12.2 |
| Sodium | 5.27 | 4.27 | 13.01 |
| Strontium | 2.38 | 1.85 | 2.67 |
| Zinc | 30.6 | 14.6 | 11.3 |
| Phytate (g/kg) | 9.79 | 7.77 | 4.93 |
| Phytate-P (g/kg) | 2.76 | 2.19 | 1.39 |
| Total P (g/kg) | 3.64 | 3.12 | 1.94 |
| Proportion (%)[a] | 75.8 | 70.2 | 71.6 |
| Grain texture Symes PSI (%)[b] | 11 | 8 | 11 |

[a] Phytate-P proportion in total P.
[b] PSI, particle size index, as described in Symes (1965), the lower value indicates harder kernel.

**Table 2.** Concentrations of total phenolic compounds, polyphenols, free, conjugated, bound and total phenolic acids and ferulic acid and RVA starch pasting profiles of Block I, HP and Liberty sorghum varieties[a].

| Polyphenolic compounds (µg/g) | Block I | HP | Liberty |
|---|---|---|---|
| Total phenolics (mg GAE/g) | 4.68 | 3.52 | 3.00 |
| Total flavonoids (µg/g) | 927.6 | 1099.5 | 648.2 |
| Anthocyanin (ABS/ml/g) | 6.53 | 5.09 | 1.30 |
| Flavan-4-ols (ABS/ml/g) | 7.98 | 4.82 | 0.84 |
| Luteolinidin | 16.07 | 1.92 | 4.39 |
| Apigeninidin | 14.75 | 5.79 | 2.13 |
| 5-methoxy luteolinidin | 7.69 | 4.19 | 4.00 |
| 7-methoxy apigeninidin | 25.29 | 3.88 | nd |
| Eriodictyol | nd | 39.5 | nd |
| Naringenin | 174.4 | 60.9 | nd |
| Luteolin | nd | 5.48 | 7.26 |
| Apigenin | 5.76 | 5.58 | 6.32 |

| Phenolic acids (µg/g) | Block I | HP | Liberty |
|---|---|---|---|
| Free phenolic acids | 35.0 | 19.5 | 12.7 |
| Free ferulic acid | 2.1 | 2.1 | 4.4 |
| Conjugated phenolic acids | 119.8 | 76.3 | 61.6 |
| Conjugated ferulic acid | 38.4 | 33.6 | 24.6 |
| Bound phenolic acids | 399.1 | 333.1 | 228.5 |
| Bound ferulic acid | 334.8 | 290.0 | 183.2 |
| Total phenolic acids | 553.8 | 426.9 | 302.8 |
| Total ferulic acid | 375.3 | 325.6 | 212.2 |
| RVA starch pasting profiles | | | |
| Peak (cP) | 2392 | 2591 | 4717 |
| Holding(cP) | 2091 | 2517 | 2810 |
| Breakdown(cP) | 300 | 74 | 1907 |
| Final(cP) | 4592 | 5554 | 5378 |
| Setback(cP) | 2501 | 3037 | 2928 |
| Peak time (minutes) | 5.63 | 6.2 | 5.17 |
| Pasting temperature (°C) | 79.9 | 82.6 | 75.9 |

[a] Analyses of phenolic compounds were conducted in triplicate whereas RVA pasting properties were measured in duplicate.

**Table 3.** Dietary composition and calculated nutrient specifications of sorghum-based diets based on Block I, HP and Liberty varieties.

| Item (g/kg) | Block I | HP | Liberty |
|---|---|---|---|
| Sorghum | 620.0 | 620.0 | 620.0 |
| Soybean meal | 224.6 | 230.0 | 225.4 |
| Canola meal | 75.0 | 75.0 | 75.0 |
| Sunflower oil | 28.0 | 20.0 | 24.0 |
| Dicalcium phosphate | 16.5 | 16.1 | 16.1 |
| Limestone | 7.4 | 6.9 | 6.9 |
| Lysine HCl | 2.5 | 3.3 | 3.5 |
| Methionine | 2.3 | 2.9 | 2.3 |
| Threonine | 0.7 | 1.2 | 1.3 |
| Arginine | 0.5 | 1.1 | 1.3 |
| Isoleucine | 0.0 | 0.3 | 0.4 |
| Valine | 0.0 | 0.6 | 0.7 |
| Sodium chloride | 1.0 | 0.7 | 0.6 |
| Vitamin-mineral premix[a] | 2.0 | 2.0 | 2.0 |
| Sodium bicarbonate | 4.5 | 4.9 | 4.9 |
| Celite | 15.0 | 15.0 | 15.0 |
| Nutrient specifications | | | |
| ME (MJ/kg) | 12.95 | 12.95 | 12.93 |
| Protein | 216.6 | 187.1 | 185.7 |
| Starch | 378.4 | 387.9 | 387.7 |
| Fat | 54.2 | 52.6 | 56.4 |
| Calcium | 7.5 | 7.5 | 7.5 |
| Total phosphorus | 7.5 | 6.9 | 6.9 |
| Available phosphorus | 3.8 | 3.8 | 3.8 |
| Lysine | 11.9 | 11.9 | 11.9 |
| Methionine | 5.6 | 5.7 | 5.7 |
| Threonine | 8.5 | 8.4 | 8.4 |
| Isoleucine | 8.7 | 8.2 | 8.2 |

| | | | |
|---|---|---|---|
| Tryptophan | 2.8 | 2.5 | 2.4 |
| Cysteine | 3.6 | 3.3 | 3.2 |
| Valine | 10.3 | 9.6 | 9.6 |
| Arginine | 11.6 | 12.7 | 11.4 |
| Histidine | 5.2 | 4.8 | 4.7 |
| Leucine | 19.9 | 17.3 | 17.1 |
| Phenylalanine | 10.9 | 9.0 | 8.9 |
| Sodium | 1.8 | 1.8 | 1.8 |
| Potassium | 7.5 | 7.6 | 7.5 |
| Chloride | 2.2 | 2.2 | 2.2 |

[a]Vitamin-trace mineral premix supplied per tonne of feed; [million international units, MIU] retinol 12, cholecalciferol 5, [g] tocopherol 50, menadione 3, thiamine 3, riboflavin 9, pyridoxine 5, cobalamin 0.025, niacin 50, pantothenate 18, folate 2, biotin 0.2, copper 20, iron 40 manganese 110, cobalt 0.25, iodine 1, molybdenum 2, zinc 90, selenium 0.3.

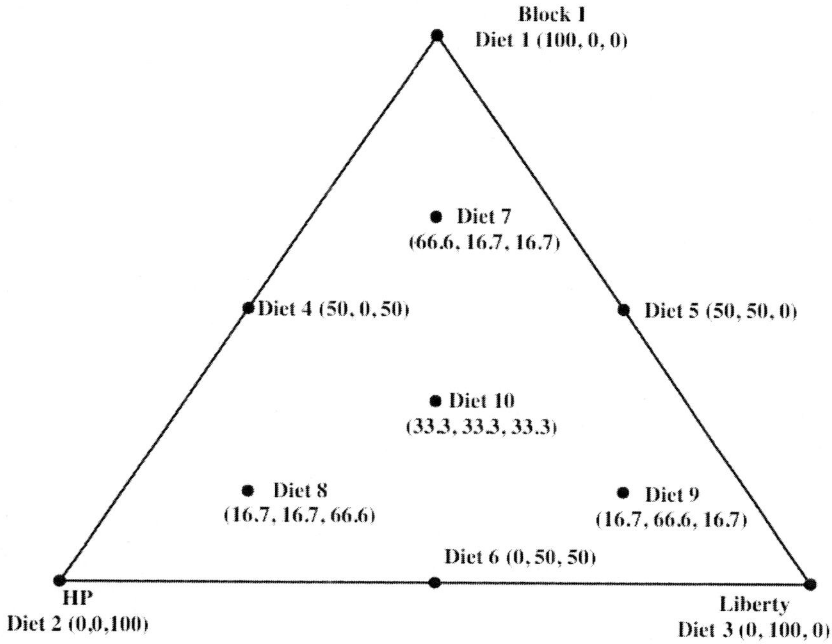


**Figure 1.** Dietary treatments in an equilateral triangle design broiler bioassay.

## 2.2. Bird management

Feather-sexed, male broiler chicks (Ross 308) were housed in an environmentally-controlled facility and were initially offered a proprietary starter ration. Birds had unlimited access to feed and water under a "23 h on–1 h off" lighting regimen for the first three days and then under a "16 h on–8 h off" lighting regimen for the rest of the study. The birds were individually identified (wing-bands) and weighed at day 7 and distributed amongst cages so that mean body weights in each cage and their variations were nearly identical. The ten dietary treatments were offered to six replicates (six birds per replicate cage) from 7 to 28 days post-hatch. Body weights were determined on days 7 and 28 and feed intakes were recorded over the entire period to calculate feed conversion ratios (FCR) with adjustments made from the weight of any dead or culled birds, which were monitored on a daily basis.

## 2.3. Sample collection and chemical analysis

Total excreta were collected from 24 to 26 days post-hatch from each cage to determine parameters of nutrient utilisation which included apparent metabolisable energy (AME), metabolisable energy to gross energy ratios (AME:GE), nitrogen (N) retention and N-corrected apparent metabolisable energy (AMEn). Excreta were dried in a forced-air oven at 80 °C for 24 h and the gross energy (GE) of excreta and diets were determined using an adiabatic bomb calorimeter. The AME values of the diets were calculated on a dry matter basis from the following equation:

$$AME_{diet} = \frac{(\text{Feed intake} \times GE_{diet}) - (\text{Excreta output} \times GE_{excreta})}{(\text{Feed intake})}$$

AME:GE Ratios were calculated by dividing AME by the GE of the appropriate diets. N contents of diets and excreta were determined using a nitrogen determinator (Leco Corporation, St Joseph, MI) and N retentions calculated from the following equation:

$$\text{Retention}(\%) = \frac{(\text{Feed intake} \times \text{Nutrient}_{diet}) - (\text{Excreta output} \times \text{Nutrient}_{excreta})}{(\text{Feed intake} \times \text{Nutrient}_{diet})} \times 100$$

N-corrected AME (AMEn MJ/kg DM) values were calculated by correcting N retention to zero using the factor of 36.54 kJ/g N retained in the body (Hill and Anderson, 1958).

The jejunum is reported to be the major site of glucose and amino acid absorption but the extent of nutrient digestion at the end of ileum is usually reported in the literature (Riesenfeld et al., 1980 and Liu and Selle, 2015). Apparent digestibility coefficients of starch and protein were determined in both distal jejunum and distal ileum in the present study. On day 28, the birds were euthanised by intravenous injection of sodium pentobarbitone and digesta samples were collected in their entirety from the two segments of small intestine (distal jejunum and distal ileum) to determine nutrient (starch and protein) digestibility coefficients and nutrient disappearance rates (g/bird/day). Acid insoluble ash (Celite®) was included in diets at 15 g/kg as an inert marker to determine starch and N digestibility. The small intestines were removed from euthanised birds and samples of digesta were gently expressed from the distal jejunum and distal ileum in their entirety and pooled for each cage. Distal jejunal samples were taken from below the mid-point between the duodenal loop and Meckel's diverticulum. Distal ileal samples were taken from below the mid-point between Meckel's diverticulum and the ileo-caecal junction. The digesta samples were freeze-dried to determine apparent digestibilities of starch and crude protein (N) using acid insoluble ash (AIA) as the inert dietary marker. About 0.65-0.85 g of dried digesta samples were used for quantifying AIA. Starch concentrations in diets and digesta were determined by a procedure based on dimethyl sulphoxide, α-amylase and amyloglucosidase as described by Mahasukhonthachat et al. (2010). N concentrations were determined as already stated and AIA concentrations were determined by the method of Siriwan et al. (1993). The apparent digestibility coefficients for starch and protein (N) in two small intestinal sites were calculated from the following equation:

$$\text{Digestibility Coefficient} = \frac{(\text{Nutrient/AIA})_{\text{diet}} - (\text{Nutrient/AIA})_{\text{digesta}}}{(\text{Nutrient/AIA})_{\text{diet}}}$$

Starch and protein (N) disappearance rates (g/bird/day) were deduced from the following equation:

$$\text{Nutrient disappearance rate}_{(\text{g/bird/day})} = \text{Feed intake}_{(\text{g/bird})} \times \text{dietary nutrient}_{(\text{g/kg})} \times \text{nutrient digestibility}_{(\text{apparentdigestibilitycoefficient})}$$

Ratios of starch to protein disappearance rates in the intestinal segments were calculated as this effectively cancels the potentially confounding influence of feed intake.

Experimental data were analysed using JMP® 9.0.0 (SAS Institute Inc. JMP Software. Cary, NC) and response surfaces were generated with R 3.0.3 software. The experimental units were replicate cage means and statistical procedures included univariate analyses of variance using the general linear models procedure, Pearson correlations and single and multiple linear regressions. A probability level of less than 5% was considered to be statistically significant. In model prediction, the non-significant coefficients were excluded which resulted in recalculation of their reduced equations for each response. When more than one models were significant, Akaike Information Criterion was used for model comparison and selection. Additionally, the response surface plots were constructed so that the effects from changing factor levels on the examined responses can be visualized. The feeding studies complied with specific guidelines approved by the Animal Ethics Committee of The University of Sydney.

# 3. RESULTS

The outcomes for growth performance, relative gizzard weights and mortality/cull rates are shown in Table 4, where there were no significant differences between the ten dietary treatments. Overall, the average weight gain was 1412 g/bird (range 1387–1441), average feed intake was 2308 g/bird (range 2238–2343) and the mean FCR was 1.634 (range 1.606–1.663). The average weight gain and feed intake was higher than 2014 Ross 308 performance objectives. The average relative gizzard weight was 21.9 g/kg and the overall 2.32% mortality rate was not related to treatment.

The nutrient utilisation results are shown in Table 5. There were significant differences ($P < 0.03$) in AME:GE ratios, which indicate the efficiency of energy utilisation between dietary treatments. The mean AME:GE ratio was 0.813, which ranged from 0.800 in treatments 1 (Block I) and 2 (HP) to 0.832 in treatment 6 (equal blend of HP and Liberty). There were no significant differences between treatments for AME, N retention and AMEn.

The effects of dietary treatments on starch digestibility coefficients and starch disappearance rates in the distal jejunum and distal ileum are shown in Table 6. Starch digestibility coefficients were not significantly different with a mean value of 0.791 (range: 0.747–0.830) in the distal jejunum and 0.888 (range: 0.867–0.910) in the distal ileum. There were, however, significant differences ($P < 0.001$) in starch disappearance rates at both sites. In the distal jejunum, starch disappearance rates ranged from 35.79 g/bird/day in treatment 4 (equal blend of HP and Block I) to 41.86 g/bird/day in treatment 6 (equal blend of HP and Liberty). In the distal ileum, starch disappearance rates ranged 35.94 g/bird/day in treatment 4 (equal blend of HP and Block I) to 44.21 g/bird/day in treatment 3 (Liberty).

The effects of dietary treatments on protein (N) digestibility coefficients and protein (N) disappearance rates in the distal jejunum and distal ileum are shown in Table 7. Protein digestibility coefficients were not significantly different with a mean value of 0.430 (range 0.378–0.509) in the distal jejunum and 0.653 (range 0.600–0.696) in the distal ileum. Protein disappearance rates were not significantly different in the distal jejunum with a mean value of 9.53 g/bird/day (range 7.60–11.18). However, there were significant differences ($P < 0.01$) in protein disappearance rates at the distal ileum which ranged from 13.14 g/bird/day in treatment 3 (Liberty), to 16.51 g/bird/day in treatment 4 (equal blend of HP and Block I).

**Table 4.** Effects of ten sorghum-based dietary treatments on growth performance, relative gizzard weights and mortality/cull rates of broiler chickens from 7 to 28 days post-hatch.

| Treatment | Weight gain (g/bird) | Feed intake (g/bird) | FCR(g/g) | Relative gizzard weight (g/kg) | Mortalities, culls (%) |
|---|---|---|---|---|---|
| 1 | 1387 | 2301 | 1.66 | 23.0 | 2.78 |
| 2 | 1431 | 2334 | 1.631 | 23.0 | 0.00 |
| 3 | 1409 | 2342 | 1.663 | 20.3 | 2.78 |
| 4 | 1421 | 2334 | 1.664 | 22.2 | 2.78 |
| 5 | 1420 | 2343 | 1.65 | 21.3 | 2.78 |
| 6 | 1441 | 2316 | 1.608 | 21.6 | 5.56 |
| 7 | 1359 | 2238 | 1.647 | 22.6 | 1.00 |
| 8 | 1432 | 2313 | 1.615 | 21.8 | 2.78 |
| 9 | 1432 | 2299 | 1.606 | 21.6 | 2.78 |
| 10 | 1392 | 2261 | 1.626 | 21.4 | 0.00 |
| SEM | 20.4 | 32.2 | 0.0183 | 0.68 | 2.256 |
| Significance (P=) | 0.139 | 0.318 | 0.142 | 0.182 | 0.714 |
| LSD (P<0.05) | - | - | - | - | - |

**Table 5.** Effects of ten sorghum-based dietary treatments on nutrient utilisation of broiler chickens from 7 to 28 days post-hatch.

| Treatment | AME (MJ/kg DM) | AME:GE ratio | N retention (%) | AMEn (MJ/kg DM) |
|---|---|---|---|---|
| 1 | 13.38 | 0.800$^{a}$ | 68.22 | 11.87 |
| 2 | 13.26 | 0.800$^{a}$ | 68.36 | 11.85 |
| 3 | 13.47 | 0.827$^{c}$ | 72.39 | 12.14 |
| 4 | 13.39 | 0.802$^{a}$ | 72.01 | 11.81 |
| 5 | 13.46 | 0.814$^{abc}$ | 71.58 | 12 |
| 6 | 13.63 | 0.832$^{c}$ | 73.41 | 12.2 |
| 7 | 13.48 | 0.807$^{ab}$ | 70.11 | 12.05 |
| 8 | 13.52 | 0.816$^{abc}$ | 71.81 | 12.03 |
| 9 | 13.54 | 0.823$^{abc}$ | 71.40 | 12.14 |
| 10 | 13.37 | 0.811$^{abc}$ | 72.12 | 11.90 |
| SEM | 0.124 | 0.0075 | 1.378 | 0.11 |
| Significance (P =) | 0.673 | 0.029 | 0.156 | 0.158 |
| LSD (P < 0.05) | - | 0.0226 | - | - |

$^{abc}$Means within columns not sharing a common suffix are significantly different ($P < 0.05$).

**Table 6.** Effects of ten sorghum-based dietary treatments on starch digestibility coefficients and starch disappearance rates (g/bird/day) in the distal jejunum and distal ileum at 28 days post-hatch.

| Treatment | Starch digestibility | | Starch disappearance | |
|---|---|---|---|---|
| | Distal jejunum | Distal ileum | Distal jejunum | Distal ileum |
| 1 | 0.767 | 0.889 | $37.81^{bc}$ | $38.02^{bc}$ |
| 2 | 0.830 | 0.894 | $41.52^{e}$ | $41.79^{de}$ |
| 3 | 0.818 | 0.910 | $43.89^{f}$ | $44.21^{f}$ |
| 4 | 0.789 | 0.891 | $35.79^{a}$ | $35.94^{a}$ |
| 5 | 0.747 | 0.860 | $37.32^{ab}$ | $37.50^{ab}$ |
| 6 | 0.786 | 0.873 | $41.86^{e}$ | $42.15^{e}$ |
| 7 | 0.801 | 0.902 | $39.13^{cd}$ | $39.40^{cd}$ |
| 8 | 0.775 | 0.867 | $39.58^{d}$ | $39.82^{d}$ |
| 9 | 0.818 | 0.900 | $39.84^{d}$ | $40.09^{d}$ |
| 10 | 0.781 | 0.894 | $39.08^{cd}$ | $39.34^{cd}$ |
| SEM | 0.0199 | 0.0194 | 0.562 | 0.556 |
| Significance (P =) | 0.123 | 0.711 | <0.001 | <0.001 |
| LSD (P < 0.05) | – | – | 1.595 | 1.579 |

[abcdef]Means within columns not sharing a common suffix are significantly different ($P < 0.05$).

**Table 7.** Effects of ten sorghum-based dietary treatments on protein (N) digestibility coefficients and protein (N) disappearance rates (g/bird/day) in the distal jejunum and distal ileum at 28 days post-hatch.

| Treatment | Protein digestibility | | Protein disappearance | |
|---|---|---|---|---|
| | Distal jejunum | Distal ileum | Distal jejunum | Distal ileum |
| 1 | 0.378 | 0.600 | 8.90 | $14.18^{abc}$ |
| 2 | 0.499 | 0.696 | 11.18 | $15.61^{cd}$ |
| 3 | 0.380 | 0.658 | 7.60 | $13.14^{a}$ |
| 4 | 0.450 | 0.692 | 10.75 | $16.51^{d}$ |
| 5 | 0.44 | 0.614 | 9.75 | $13.60^{ab}$ |
| 6 | 0.398 | 0.638 | 8.48 | $13.54^{ab}$ |
| 7 | 0.373 | 0.637 | 8.34 | $14.27^{abc}$ |
| 8 | 0.419 | 0.639 | 9.47 | $14.50^{abc}$ |
| 9 | 0.509 | 0.674 | 10.80 | $14.34^{abc}$ |
| 10 | 0.453 | 0.681 | 10.05 | $15.11^{bcd}$ |
| SEM | 0.0408 | 0.0258 | 0.886 | 0.602 |
| Significance (P =) | 0.198 | 0.108 | 0.09 | 0.008 |
| LSD (P< 0.05) | – | – | – | 1.709 |

$^{abcd}$ Means within columns not sharing a common suffix are significantly different ($P < 0.05$).

The optimal ratios of the three sorghum-based diets for any bird performance parameter may be calculated from the contour plots generated by the ten dietary treatments. For example, the relationship between proportions of the three basal diets and 7–28 day weight gain weight gain is predicted by the following equation ($r^2 = 0.99$, $P < 0.001$, lack of fit $P = 0.241$). In this equation, $PBlock\ I$,$P_{HP}$ and $P_{Liberty}$ represent the percentage of diets based on sorghum Block I, HP and Liberty.

$$Gain = 13.792P_{BlockI} + 14.377P_{HP} + 14.173P_{Liberty} - 0.003158P_{BlockI}P_{HP} + 0.001309P_{BlockI}P_{Liberty}$$

As illustrated in Fig. 2(a), increasing the inclusion level of diet based on Block I compromised weight gain and the predicted optimal weight gain of 1439 g/bird would be achieved with a diet based on 79.2% HP and 20.8% Liberty but 0.0% Block I. Similarly, the most efficient FCR can be depicted as shown in Fig. 2(b). The relationship between FCR and the proportions of the three basal diets may be predicted by the significant following ($r^2 = 0.99$, $P < 0.001$, lack of fit $P = 0.801$) equation:

$$FCR = 0.01655P_{BlockI} + 0.01631P_{HP} + 0.01651P_{Liberty} - 1.783 \times 10^{-5}P_{HP}P_{Liberty}$$

Clearly, higher inclusion rates of Block I compromised feed efficiency and the predicted best FCR 1.486 g/g was achieved when the proportions of diets based on Block I, HP and Liberty equal to 0.0%, 55.6% and 44.4%, respectively.

In terms of energy utilisation, as illustrated in Fig. 3(a), the relationship between AMEn and the proportions of three basal diets was predicted by the following equation ($r^2 = 0.99$, $P < 0.001$, lack of fit $P = 0.618$):

$$AMEn = 0.01191P_{BlockI} + 0.1186P_{HP} + 0.1214P_{Liberty} - 1.927 \times 10^{-5}P_{BlockI}P_{HP} - 4.226 \times 10^{-6}P_{BlockI}P_{Liberty} + 7.799 \times 10^{-5}P_{HP}P_{Liberty}$$

Increasing the inclusion rates of diets based on Block I and HP compromised AMEn and the predicted highest AMEn 12.38 MJ/kg was achieved when the proportions of diets based on Block I, HP and Liberty were equal to 0.0%, 32.0% and 68.0%, respectively.

Significant differences were confined to AME:GE ratios. Then, as shown in Fig. 3(b), the relationship between AME:GE ratios and the proportions of the three basal diets may be predicted by the following significant ($r^2 = 0.99$, $P < 0.001$, lack of fit $P = 0.992$) equation:

$$ME{:}GE = 0.008P_{BlockI} + 0.008016P_{HP} + 0.008262P_{Liberty} + 6.520 \times 10^{-6}P_{HP}P_{Liberty}$$

The predicted maximum AME:GE ratio of 0.845 MJ/MJ will be achieved when the proportions of diets based on Block I, HP and Liberty equal to 0.0%, 31.1% and 68.9%, respectively. Thus, essentially, the most efficient energy utilisation would be generated with diets based on a 69:31 blend of Liberty and HP.

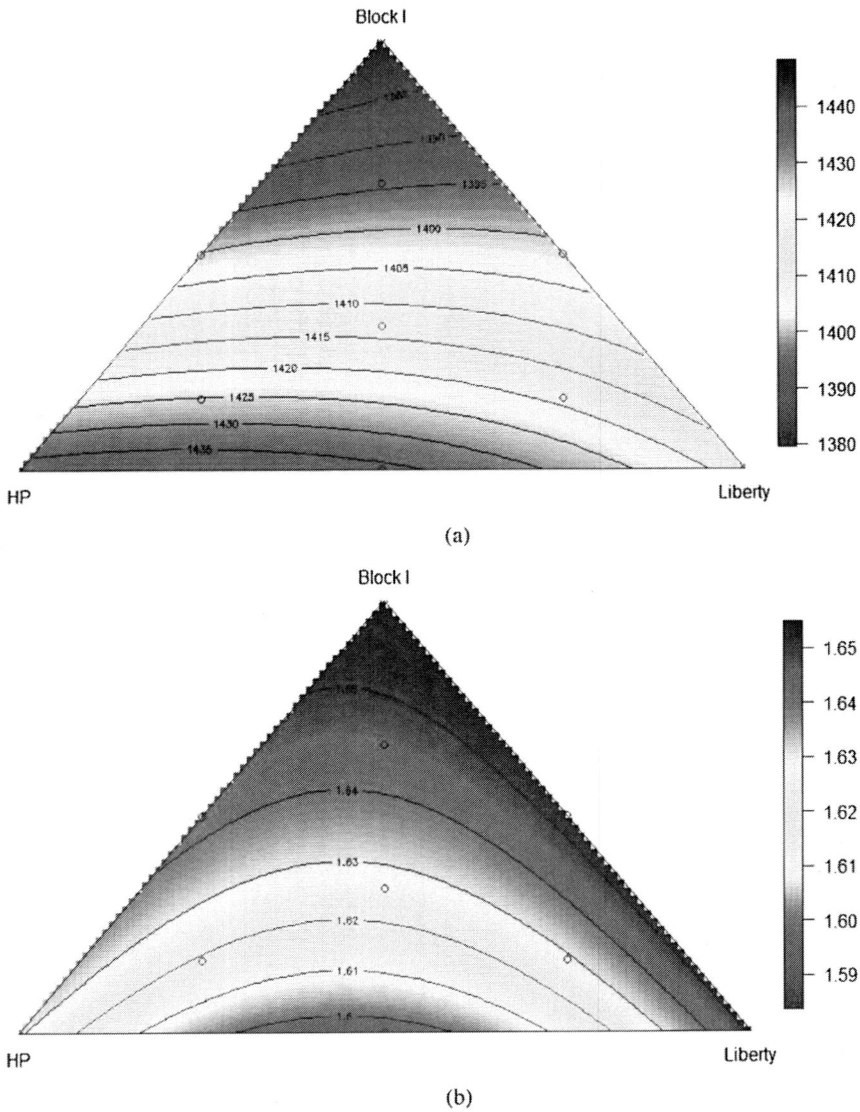


**Figure 2.** Response surfaces of the relationships between the proportions of three basal diets and weight gain (a), feed conversion ratio (b).

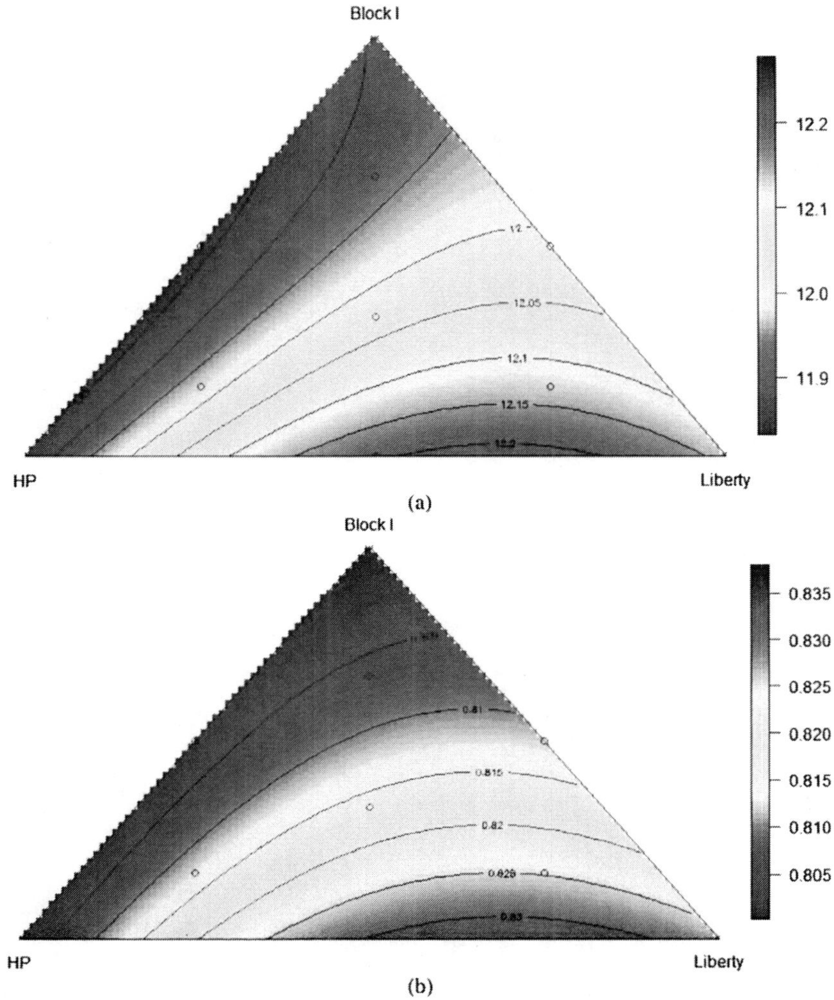


**Figure 3.** Response surfaces of the relationships between the proportions of three basal diets and nitrogen-corrected apparent metabolisable energy (a) and the ratio of metabolisable energy to gross energy (b).

There were multiple linear relationships ($r^2 = 0.99$, $P < 0.001$, lack of fit $P > 0.50$) between the proportion of three sorghums and distal ileal digestibility coefficients of starch and protein (Fig. 4). The relationship between protein digestibility and the proportions of the three basal diets may be predicted by the following equation:

Distal ileal protein digestibility $= 0.00618P_{BlockI} + 0.00691P_{HP} + 0.00651P_{Liberty}$

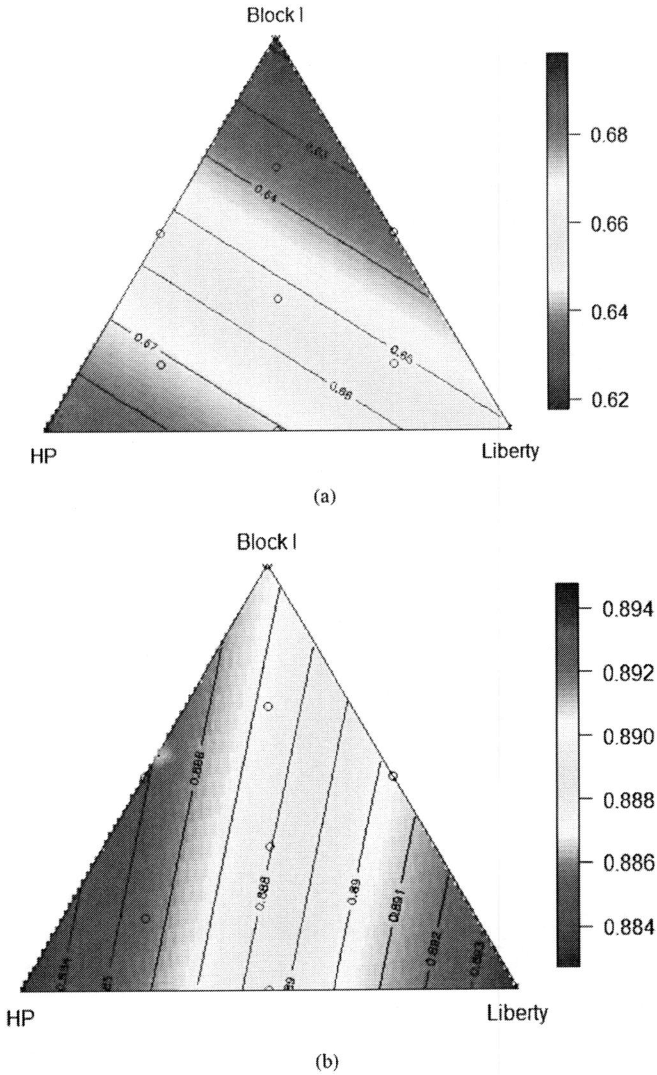


**Figure 4.** Response surfaces of the relationships between the proportions of three basal diets and apparent digestibility coefficients of protein (a) and starch (b) in the distal ileum.

The predicted maximum distal ileal protein digestibility of 0.691 will be achieved when the diet contain 100% sorghum HP.

In contrast, the relationship between starch digestibility and the proportions of the three basal diets may be predicted by the following equation:

Distal ileal starch digestibility = $0.00887P_{BlockI} + 0.00883P_{HP} + 0.00894P_{Liberty}$

The predicted optimal distal ileal starch digestibility of 0.894 will be achieved when the diet contain 100% Liberty sorghum.

The concentrations of kafirin, total phenolics and phytate in the three sorghum varieties have been established and therefore, the concentrations of these three components in the ten dietary treatments derived from the sorghum component can be calculated. In this context, kafirin can only be derived from sorghum. Phenolic compounds are present in soybean meal and canola meal but grain sorghum is a particularly rich source of phenolics (Bravo, 1998). Phytate is also present in soybean meal and canola meal at concentrations in the order of 13.8 g/kg and 22.9 g/kg, respectively (Selle and Ravindran, 2007). Nevertheless, concentrations of phenolics and phytate derived from the two protein meals would remain essentially constant across the dietary treatments and variations in these concentrations would largely stem from the sorghum component. On this basis, the calculated concentrations of kafirin, phytate, total and individual phenolic compounds in the ten experimental diets based on the three sorghums were tabulated inTable 8. The 3-deoxyanthocyanins are mainly concentrated in the grain pericarp and luteolinidin, apigeninidin, 5-methoxy-luteolinidin and 7-methoxy-apigeninidin were detected as the four major polyphenols within this class (Awika et al., 2004 and Khoddami et al., 2015); therefore, the total concentrations of 3-deoxyanthocyanins were reported inTable 8.

Table 9 summarised the relationships of kafirin, total phenolics and phytate weight growth performance and nutrient utilisations. Concentrations of kafirin, total phenolics and phytate in the diets were correlated with starch disappearance rates in the distal jejunum and distal ileum ($P < 0.02$); however, they were not correlated with disappearance rates of protein (N). In addition to Table 9, weight gain was negatively correlated ($r = -0.680$, $P = 0.031$) to concentrations of 3-deoxyanthocyanins in the diets (Fig. 5). Feed conversion ratio tended to correlated with dietary concentrations of luteolinidin ($r = 0.587$, $P = 0.074$) and 5-methoxy luteolinidin ($r = 0.572$, $P = 0.084$). Protein disappearance rate in the distal ileum was negatively correlated to free ferulic acid in the diets ($r = -0.710$, $P = 0.021$) as shown in Fig. 6.

**Table 8.** Calculated concentrations of kafirin, phytate, total phenolics, anthocyanin, flavan-4-ols and 3-deoxyanthocyanins compounds in the ten experimental diets based on sorghum Block I, HP and Liberty.

| Diet | Kafirin (g/kg) | Phytate (g/kg) | Total phenolics (mg GAE/g) | Total favonoids (µg/g) | Flavan-4-ols (ABS/ml/g) | 3-Deoxyanthocyanins (µg/g) | Free ferulic acid (µg/g) |
|---|---|---|---|---|---|---|---|
| 1 | 67.1 | 9.79 | 4.68 | 928 | 7.98 | 63.80 | 2.1 |
| 2 | 50.5 | 7.77 | 3.52 | 1100 | 4.82 | 15.78 | 2.1 |
| 3 | 41.4 | 4.93 | 3.00 | 648 | 0.84 | 10.91 | 4.4 |
| 4 | 58.8 | 8.78 | 4.10 | 1014 | 6.40 | 39.80 | 2.1 |
| 5 | 54.3 | 7.36 | 3.84 | 788 | 4.41 | 37.36 | 3.3 |
| 6 | 46 | 6.35 | 3.26 | 874 | 2.83 | 13.35 | 3.3 |
| 7 | 60 | 8.64 | 4.21 | 910 | 6.26 | 46.96 | 2.5 |
| 8 | 49.7 | 7.34 | 3.49 | 968 | 4.44 | 21.06 | 2.4 |
| 9 | 45.2 | 5.92 | 3.23 | 742 | 2.46 | 18.65 | 3.6 |
| 10 | 52.9 | 7.49 | 3.73 | 891 | 4.54 | 30.13 | 2.9 |

**Table 9.** Summary of correlations between concentrations of kafirin, total phenolics and phytate in the diets and growth performance and nutrient utilisations in broiler chickens.

| | | Weight gain | FCR | AMEn | AME:GE | N retention | Starch disappearance rates | | Protein (N) disappearance rates | |
|---|---|---|---|---|---|---|---|---|---|---|
| | | | | | | | Distal jejunum | Distal ileum | Distal jejunum | Distal ileum |
| Kafirin | r = | −0.607 | 0.504 | −0.658 | −0.801 | −0.589 | −0.759 | −0.757 | 0.072 | 0.353 |
| | P-value | *0.063* | *0.138* | *0.039* | *0.005* | *0.074* | *0.011* | *0.011* | *0.843* | *0.318* |
| Phytate | r = | −0.485 | 0.388 | −0.633 | −0.869 | −0.626 | −0.737 | −0.736 | 0.216 | 0.516 |
| | P-value | *0.156* | *0.268* | *0.049* | *0.001* | *0.053* | *0.015* | *0.015* | *0.548* | *0.127* |
| Total phenolics | r = | −0.629 | 0.521 | −0.744 | −0.784 | −0.578 | −0.756 | −0.754 | 0.044 | 0.320 |
| | P-value | *0.051* | *0.123* | *0.014* | *0.007* | *0.080* | *0.011* | *0.012* | *0.904* | *0.367* |

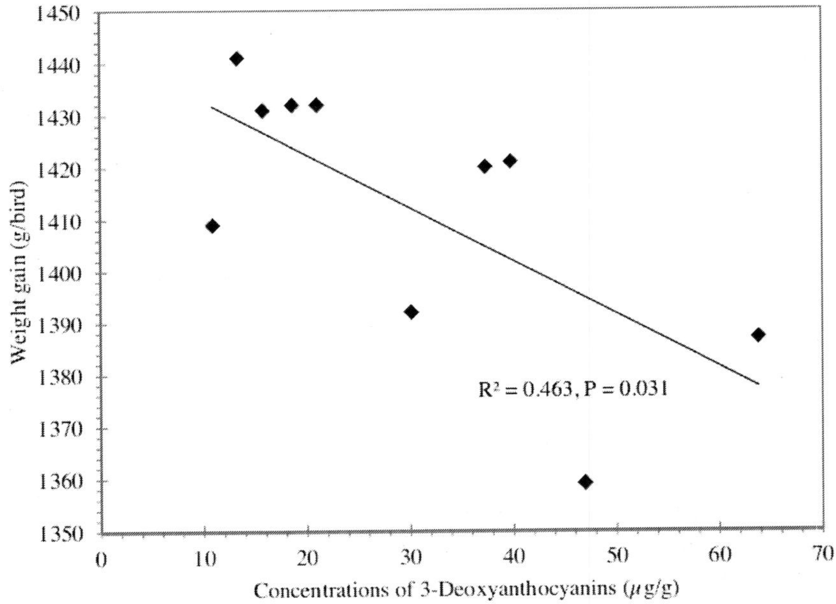


**Figure 5.** The negative linear regressions between weight gain and concentrations of 3-deoxyanthocyanins.

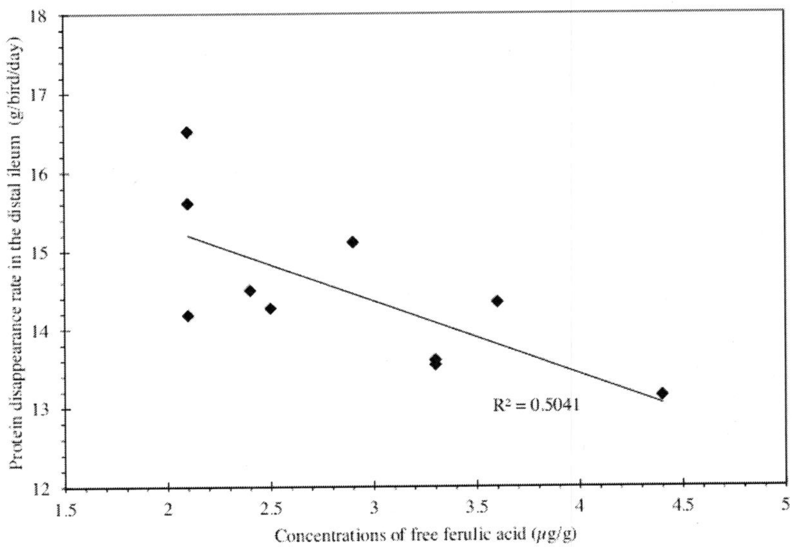


**Figure 6.** The negative linear regressions between protein disappearance rate in the distal ileum was negatively correlated to free ferulic acid in the diets.

# 4. DISCUSSION

The blends of Liberty and HP to achieve optimal weight gain, FCR and AME:GE are somewhat different but in all instances Block I would be discarded. Phytate and phenolic acids, including ferulic acid, are not unique to sorghum but high levels of polyphenols (Bravo, 1998) and kafirin protein bodies distinguish sorghum from other feed grains. In the present study, Block I sorghum contained 33% more kafirin (67.1 versus 50.5 g/kg), 26% more phenolic compounds (4.68 versus 3.52 mg GAE/g), 30% more phenolic acids (554 versus 427 μg/g) and 26% more phytate (9.79 versus 7.77 g/kg) than HP sorghum. Block I also contained 62% more kafirin (67.1 versus 41.4 g/kg), 56% more phenolic compounds (4.68 versus 3.00 mg GAE/g), 83% more phenolic acids (554 versus 303 μg/g) and 99% more phytate (9.79 versus 4.93 g/kg) than Liberty sorghum. Consequently it is not surprising that Block I sorghum was effectively rejected by the triangular contour plots for weight gains, feed conversion ratios and AME:GE ratios or the efficiency of energy utilisation.

Condensed tannin is a polyphenolic phytochemical with pronounced anti-nutritive properties in poultry (Nyachoti et al., 1997); however, Australian sorghum crops do not presently contain condensed tannin (Khoddami et al., 2015). Nevertheless, as reported by Khoddami et al. (2015), sorghum crops contain a diverse variety of polyphenolic compounds and phenolic acids, which are usually at substantially higher concentrations than other feed grains (Mangan, 1988 and Bravo, 1998). It is our contention is that the anti-nutritive properties of phenolic compounds in grain sorghum are not confined to condensed tannin but extend to other 'non-tannin' phenols (Liu et al., 2015a). In the present study, weight gain was negatively correlated ($r = -0.680$, $P = 0.031$) to concentrations of 3-deoxyanthocyanins in the diets and FCR tended to correlated with dietary concentrations of luteolinidin ($r = 0.587$, $P = 0.074$) and 5-methoxy luteolinidin ($r = 0.572$, $P = 0.084$). Moreover, protein disappearance rate in the distal ileum was negatively correlated to free ferulic acid in the diets ($r = -0.710$, $P = 0.021$). Duodu et al. (2003) suggested that phenolic compounds, including phenolic acids and flavenoids, contain hydroxyl groups and may interact with and form complexes with proteins. The 3-deoxyanthocyanins are mainly concentrated in the grain pericarp and pigmented sorghums are a rich source of unique 3-deoxyanthocyanidins that give yellow to orange color in acidic media (Awika et al., 2004). In Australia, it has been reported that broiler chickens offered white-sorghum-based diets had better performance than chickens offered red-sorghum-based diets (Liu et al., 2013 and Selle et al., 2013). In line with the present study, Rooney et al.

(2007) showed that tannin-free red sorghum had higher total phenols than white sorghum. Liberty is the only breed of white variety sorghum in Australia and it contained less phytate and kafirin than red sorghums in the present study. Nevertheless, the negative impact of non-tannin phenolics on performance is evident in the present study.

As shown in Table 9, energy utilisation (AMEn and AME:GE ratios) and starch disappearance rates in the distal ileum were negatively influenced by kafirin, total phenolic compounds and phytate concentrations. Multiple linear regression was conducted to compare the relative importance of kafirin, total phenolic compounds and phytate on energy utilisation and they could not be considered separately; therefore, they may be equally important. Kafirin is present in sorghum endosperm as discrete protein bodies with a central core of α-kafirin and a periphery of β- and γ-kafirin fractions. The protein bodies are embedded in the glutelin protein matrix in close association with starch granules (Selle et al., 2010a). This close proximity of kafirin protein bodies and starch granules in sorghum endosperm presumably facilitates any physical and chemical starch-protein interactions which are considered to be important (Rooney and Pflugfelder, 1986), if not clearly defined (Truong et al., 2015). While the underlying mechanisms have yet to be properly clarified, it has been reported that kafirin has the capacity to compromise starch/energy utilisation in grain sorghum in poultry (Salinas et al., 2006). This may be a consequence of disuphide cross-linkages, especially in the β- and γ-kafirin periphery of protein bodies, as Ezeogu et al. (2008) attributed the low starch digestibility of cooked sorghum flour to a more disulfide-bonded protein matrix limiting the expansion of the starch granules and hence amylase access. Moreover, Hamaker and Bugusu (2003) concluded that sorghum proteins, especially kafirin, negatively influence starch gelatinisation and the extent of starch digestion.

The consideration of the interactions between starch and phenolic compounds have been reviewed by Zhu (2015) and Hung et al. (2013) reported that ferulic acid formed complexes with debranched starch ranging from 6.8 to 31.5 mg ferulic acid/g starch. Moreover, Kim et al. (2000) reported that luteolin has the capacity to inhibit α-glucosidase and amylase and (Kottra and Daniel, 2007) found that flavonoids inhibit glucose and sodium uptakes via SGLT1 transporters. Earlier, Welsch et al. (1989) had reported that phenolic compounds, including ferulic acid, inhibit $Na^+$-dependent intestinal glucose uptakes in rats. This was attributed to the dissipation of the $Na^+$ electrochemical gradient in the gut mucosa, which drives the active co-absorption of glucose and Na. Phytate also has the capacity to inhibit amylase (Sharma et al., 1978) and depress the activity of $Na^+/K^+$-ATPase or the 'sodium pump' (Dilworth et al., 2005), which drives the co-absorption of glucose and sodium. It has been suggested that phenolic compounds and phytate share analagous anti-nutritive properties (Selle et al., 2010b) and it appears that these two factors in tandem could compromise the digestion of sorghum starch and, probably more importantly, the subsequent absoprtion of glucose.

The influence of concentrations of kafirin, total phenolic compounds, phytate on RVA pasting properties in sorghum *per se* is tabulated in Table 10. Liu et al. (2014) suggested that although the cooking and cooling during Rapid Visco-Analyses of starch does not duplicate conditions in the gastrointestinal tract of poultry, it should provide information on the effects of the various forms of grain processing on starch properties. Moreover,Doucet et al. (2010) found that final starch viscosity of diets predicted starch digestibility coefficients determined at the small intestinal mid-point in weaner pigs. Therefore, it is possible to consider Rapid Visco-Analyses as a rapid *in vitro* assessment of starch properties in cereal grains. In the present study, the concentrations of kafirin, total phenolic compounds, phytate were correlated with RVA Peak, Holding and Final viscosity. This again suggested kafirin, total phenolic compounds, phytate may influence functional properties of starch in sorghum grain.

It is noteworthy that sorghums with low protein concentrations may benefit growth performance of broiler chickens (Selle et al., 2010a). Taylor et al. (1984) surveyed 44 grain sorghum varieties; the average protein content was 112 g/kg with proportions of 47.8% kafirin and 24.7% glutelin. Interestingly, the kafirin proportion increased at the expense of glutelin with increasing sorghum protein contents. Kafirin contains low concentrations of basic amino acids including arginine, histidine and especially lysine and the digestibility of amino acids in kafirin is inferior to the 'non-kafirin' component of sorghum protein (Shull et al., 1992, Watterson et al., 1993 and Selle et al., 2010a). It is probable that high protein sorghums contain higher proportions of kafirin and lower amino acid digestibilities than low protein sorghums. Moreover, Liu and Selle (2015) considered the importance of digestive dynamics of starch and protein; protein digestion in the proximal small intestine may be more important for both protein utilisation and feed conversion efficiency. Therefore, both the extent and rate of protein digestion are pivotal to growth performance in broiler chickens. In order to formulate nutritionally adequate diets, diets based on low protein sorghums require higher inclusions of other protein sources, including soybean meal and synthetic amino acids. In practice, broiler diets are formulated to contain adequate digestible amino acid concentrations rather than total amino acids. Therefore, diets based on low protein sorghums contain higher levels of soybean meal and synthetic amino acids and have more rapid rates of protein digestion which should benefit feed conversion efficiency.

**Table 10.** Pearson correlations between concentrations of kafirin, total phenolic compounds, phytate in grain sorghum and peak, holding and final RVA starch viscosities.

| | Kafirin | Phenols | Phytate | Peak | Holding | Final |
|---|---|---|---|---|---|---|
| Kafirin | 1.000 | | | | | |
| Phenols | r = 0.999<br>P < 0.001 | 1.000 | | | | |
| Phytate | r = 0.965<br>P < 0.001 | r = 0.951<br>P < 0.001 | 1.000 | | | |
| Peak | r = -0.817<br>P = 0.004 | r = -0.787<br>P = 0.007 | r = -0.939<br>P < 0.001 | 1.000 | | |
| Holding | r = -0.998<br>P < 0.001 | r = -0.994<br>P < 0.001 | r = -0.980<br>P < 0.001 | r = 0.851<br>P < 0.002 | 1.000 | |
| Final | r = -0.863<br>P = 0.001 | r = -0.888<br>P = 0.001 | r = -0.702<br>P = 0.024 | r = 0.415<br>P = 0.234 | r = 0.831<br>P < 0.003 | 1.000 |

# 5. CONCLUSIONS

In conclusion, the present study indicates that a model sorghum variety for chicken-meat production would be relatively high in starch but low in protein. A low-protein sorghum would automatically contain less kafirin and would be enhanced further if the kafirin proportion of protein was also reduced. Ideally this model sorghum would also contain relatively low levels of polyphenols and phenolic acids. While phytate, possibly in tandem with phenolic compounds, is a negative in the overall equation, the deleterious effects of sorghum phytate can be counteracted by the inclusion of exogenous phytases in sorghum-based diets, which is a routine procedure. Indeed, more robust phytase responses in broiler chickens well may be observed when they are offered diets based on such a model sorghum simply because phytase cannot ameliorate the anti-nutritive properties of kafirin and phenolic compounds.

# CONFLICT OF INTEREST

The authors declare no conflicts of interest.

# ACKNOWLEDGMENTS

We would particularly like to thank Dr Karlie Nielsen for quantifying kafirin in sorghum while at the Australian Proteome Analysis Facility (Macquarie University) and Dr David Cadogan (Feedworks) for formulating the diets. We would like to acknowledge the Chicken-meat Committee of the Rural Industries Research and Development Corporation (RIRDC) for their ongoing encouragement and financial support (PRJ-008695) and thePoultry Cooperative Research Centre for their support of Ms Ha Truong's PhD candidature.

# REFERENCES


1. Awika, J.M., Rooney, L.W., Waniska, R.D., 2004. Properties of 3-deoxyanthocyanins from sorghum. J. Agric. Food Chem. 52, 4388–4394.

2. Black, J.L., Hughes, R.J., Nielsen, S.G., Tredrea, A.M., MacAlpine, R., Van Barneveld, R.J., 2005. The energy value of cereal grains, particularly wheat and sorghum, for poultry. Proc. Aust. Poult. Sci. Symp. 17, 21–29.

3. Bravo, L., 1998. Polyphenols: Chemistry, dietary sources, metabolism, and nutritional significance. Nutr. Rev. 56, 317–333.

4. Cohen, S.A., Michaud, D.P., 1993. Synthesis of a fluorescent derivatizing reagent, 6-aminoquinolyl-n-hydroxysuccinimidyl carbamate, and its application for the analysis of hydrolysate amino-acids via high-performance liquid-chromatography. Anal. Biochem. 211, 279–287.

5. Dilworth, L.L., Omoruyi, F.O., Asemota, H.N., 2005. Digestive and absorptive enzymes in rats fed phytic acid extract from sweet potato (Ipomoea batatas). Diabetol. Croat. 34, 59–65.

6. Doucet, F.J., White, G.A., Wulfert, F., Hill, S.E., Wiseman, J., 2010. Predicting in vivo starch digestibility coefficients in newly weaned piglets from in vitro assessment of diets using multivariate analysis. Br. J. Nutr. 103, 1309–1318.

7. Duodu, K.G., Taylor, J.R.N., Belton, P.S., Hamaker, B.R., 2003. Factors affecting sorghum protein digestibility. J. Cereal Sci. 38, 117–131.

8. Ezeogu, L.I., Duodu, K.G., Emmambux, M.N., Taylor, J.R.N., 2008. Influence of cooking conditions on the protein matrix of sorghum and maize endosperm flours. Cereal Chem. 85, 397–402.

9. Hamaker, B.R., Bugusu B.A., 2003. Overview: sorghum proteins and food quality. In: Afripro Workshop on the Proteins of Sorghum and Millets: Enhancing Nutritional and Functional Properties for Africa. Pretoria, South Africa, 2–4 April 2003.

10. Hamaker, B.R., Mohamed, A.A., Habben, J.E., Huang, C.P., Larkins, B.A., 1995. Efficient procedure for extracting maize and sorghum

kernel proteins reveals higher prolamin contents than the conventional method. Cereal Chem. 72, 583–588.

11. Hill, F.W., Anderson, D.L., 1958. Comparison of metabolizable energy and productive energy determinations with growing chicks. J. Nutr. 64, 587–603.

12. Hung, P.V., Phat, N.H., Phi, N.T.L., 2013. Physicochemical properties and antioxidant capacity of debranched starch-ferulic acid complexes. Starch-Starke 65, 382–389.

13. Khoddami, A., Truong, H.H., Liu, S.Y., Roberts, T.H., Selle, P.H., 2015. Concentrations of specific phenolic compounds in six red sorghums influence nutrient utilisation in broiler chickens. Anim. Feed Sci. Technol. 210, 190–199.

14. Kim, J.S., Kwon, C.S., Son, K.H., 2000. Inhibition of alpha-glucosidase and amylase by luteolin, a flavonoid. Biosci. Biotechnol. Biochem. 64, 2458–2461.

15. Kottra, G., Daniel, H., 2007. Flavonoid glycosides are not transported by the human Na+/glucose transporter when expressed in Xenopus laevis oocytes, but effectively inhibit electrogenic glucose uptake. J. Pharmacol. Exp. Ther. 322, 829–835.

16. Liu, S.Y., Selle, P.H., 2015. A consideration of starch and protein digestive dynamics in chicken-meat production. World Poult. Sci. J. 71, 297–310.

17. Liu, S.Y., Selle, P.H., Cowieson, A.J., 2013. Influence of white- and red-sorghum varieties and hydrothermal component of steam-pelleting on digestibility coefficients of amino acids and kinetics of amino acids, nitrogen and starch digestion in diets for broiler chickens. Anim. Feed Sci. Technol. 186, 53–63.

18. Liu, S.Y., Selle, P.H., Khoddami, A., Roberts, T.H., Cowieson, A.J., 2014. Graded inclusions of sodium metabisulphite in sorghum-based diets: II. Modification of starch pasting properties in vitro and beneficial impacts on starch digestion dynamics in broiler chickens. Anim. Feed Sci. Technol. 190, 68–78.

19. Liu, S.Y., Fox, G., Khoddami, A., Neilson, K.A., Truong, H.H., Moss, A.F., Selle, P.H., 2015a. Grain sorghum: a conundrum for chicken-meat production. Agriculture 5, 1224–1251.

20. Liu, S.Y., Selle, P.H., Simpson, S.J., Cowieson, A.J., Raubenheimer, D., 2015b. The euclidian pathway to more instructive broiler bioassays: nutritional geometry. Proc. Aust. Poult. Sci. Symp. 26, 52–60.

21. Mahasukhonthachat, K., Sopade, P.A., Gidley, M.J., 2010. Kinetics of starch digestion and functional properties of twin-screw extruded sorghum. J. Cereal Sci. 51, 392–401.

22. Mangan, J.L., 1988. Nutritional effects of tannins in animal feeds. Nutr. Res. Rev. 1, 209–231.

23. Nyachoti, C.M., Atkinson, J.L., Leeson, S., 1997. Sorghum tannins: a review. World Poult. Sci. J. 53, 5–21.

24. Raubenheimer, D., 2011. Toward a quantitative nutritional ecology: the right-angled mixture triangle. Ecol. Monogr. 81, 407–427.

25. Riesenfeld, G., Sklan, D., Bar, A., Eisner, U., Hurwitz, S., 1980. Glucose-absorption and starch digestion in the intestine of the chicken. J. Nutr. 110, 117–121.

26. Rooney, L.W., Pflugfelder, R.L., 1986. Factors affecting starch digestibility with special emphasis on sorghum and corn. J. Anim. Sci. 63, 1607–1623.

27. Rooney, W.L., Blumenthal, J., Bean, B., Mullet, J.E., 2007. Designing sorghum as a dedicated bioenergy feedstock. Biofuels Bioprod. Biorefin. 1, 147–157.

28. Salinas, I., Pro, A., Salinas, Y., Sosa, E., Becerril, C.M., Cuca, M., Cervantes, M., Gallegos, J., 2006. Compositional variation amongst sorghum hybrids: effect of kafirin concentration on metabolizable energy. J. Cereal Sci. 44, 342–346.

29. Selle, P.H., Ravindran, V., 2007. Microbial phytase in poultry nutrition. Anim. Feed Sci. Technol. 135, 1–41.

30. Selle, P.H., Cadogan, D.J., Li, X., Bryden, W.L., 2010a. Implications of sorghum in broiler chicken nutrition. Anim. Feed Sci. Technol. 156, 57–74.

31. Selle, P.H., Ravindran, V., Cowieson, A.J., Bedford, M.R., 2010b. Phytate and Phytase. Cabi Publishing-C a B Int., Cabi Publishing, Wallingford 0x10 8de, Oxon, Uk.

32. Selle, P.H., Liu, S.Y., Cai, J., Cowieson, A.J., 2013. Steam-pelleting temperatures, grain variety, feed form and protease supplementation of mediumly-ground sorghum-based broiler diets: influences on growth performance, relative gizzard weights, nutrient utilisation, starch and nitrogen digestibility. Anim. Prod. Sci. 53, 378–387.

33. Sharma, C.B., Goel, M., Irshad, M., 1978. Myoinositol hexaphosphate as a potential inhibitor of alpha-amylases. Phytochemistry 17, 201–204.

34. Shull, J.M., Watterson, J.J., Kirleis, A.W., 1992. Purification and immunocytochemical localization of kafirins in sorghum-bicolor (l-moench) endosperm. Protoplasma 171, 64–74.

35. Siriwan, P., Bryden, W.L., Mollah, Y., Annison, E.F., 1993. Measurement of endogenous amino-acid losses in poultry. Br. Poult. Sci. 34, 939–949.

36. Skoglund, E., Carlsson, N.G., Sandberg, A.S., 1997a. Analysis of inositol mono- and diphosphate isomers using high-performance ion chromatography and pulsed amperometric detection. J. Agric. Food Chem. 45, 4668–4673.

37. Skoglund, E., Carlsson, N.G., Sandberg, A.S., 1997b. Determination of isomers of inositol mono- to hexaphosphates in selected foods and intestinal contents using high-performance ion chromatography. J. Agric. Food Chem. 45, 431–436.

38. Symes, K.J., 1965. Inheritance of grain hardness in wheat as measured by particle size index. Aust. J. Agric. Res. 16, 113–123.

39. Taylor, J.R.N., Schussler, L., Vanderwalt, W.H., 1984. Fractionation of proteins from low-tannin sorghum grain. J. Agric. Food Chem. 32, 149–154.

40. Truong, H.H., Liu, S.Y., Selle, P.H., 2015. Starch utilisation in chicken-meat production: the foremost influential factors. Anim. Prod. Sci.

41. Wallace, J.C., Lopes, M.A., Paiva, E., Larkins, B.A., 1990. New methods for extraction and quantitation of zeins reveal a high content of gamma-zein in modified opaque-2 maize. Plant Physiol. 92, 191–196.

42. Watterson, J.J., Shull, J.M., Kirleis, A.W., 1993. Quantitation of alpha-karafins beta-karafins, and gamma-kafirins in vitreous and opaque endosperm of sorghum-bicolor. Cereal Chem. 70, 452–457.

43. Welsch, C.A., Lachance, P.A., Wasserman, B.P., 1989. Dietary phenolic-compounds – inhibition of Na+-dependent D-glucose uptake in rat intestinal brush-border membrane-vesicles. J. Nutr. 119, 1698–1704.

44. Zhu, F., 2015. Interactions between starch and phenolic compound. Trends Food Sci. Tech. 43, 129–143.

# CHAPTER 3

# Nutrition and Health of Dairy Animals

**Cazim Crnkic[1] and Aida Hodzic[1]**

[1] University of Sarajevo, Veterinary faculty, Bosnia & Herzegovina

## 1. INTRODUCTION

Modern breeds of dairy animals are able to produce huge amount of milk. In attempt to consume, digest and metabolize enough nutrients to satisfy lactation needs, those animals are exposed to serious stress conditions that can affect their health. Health problems which arise from those conditions are mainly related to impaired ability to metabolize enough nutrients to compensate for those lost in milk. They are known as metabolic or production diseases and may be of great economic importance in milk production systems.

Although metabolic diseases have become a common problem on dairy farms, they still require a serious attention to be controlled. The incidences of these disorders can be reduced by proper nutrition of animals. Also, some of the specific strategies in feeding practice offer additional advantages in prevention of nutrition-related metabolic diseases.

# 2. NUTRITION-RELATED METABOLIC DISORDERS, THEIR ETIOLOGY AND CONSEQUENCES

The most important metabolic diseases in dairy cows, ewes and goats can be discussed as energy and/or fiber related, lipid related or vitamin and mineral related disorders. However, they are not easily categorized according to their cause, thus the pathogenesis and hallmarks of each disease should be considered for their categorization. Energy related disorders (related to energy density of the diet or feed intake) include fatty liver and ketosis, rumen acidosis, laminitis, displaced abomasums and milk fat depression. As energy density and fiber content of the diets are often inversely related, most of these diseases can be considered as fiber related too. Fatty liver and ketosis also can be categorized as lipid related disorders due to changes in lipid metabolism in affected animals (Grummer, 1993). The most important mineral/vitamin related disorders are hypocalcaemia, hypomagnesaemia, udder edema, retained placenta and metritis. Not all of them are ultimately caused by mineral composition of the diet, but may be prevented by manipulation of minerals or vitamins in the diet. Broken homeostatic mechanisms often become major etiological factors in the development of metabolic disease, and consequences varies from reduced production and impaired reproductive performances to increased risk to develop other diseases.

## 2.1. Fatty Liver

Fatty liver (hepatic lipidosis, fat cow syndrome) is a metabolic disorder characterized by a high content of lipids and triglycerides in the liver (Ingvartsen, 2006). The disease occurs in periparturient period, primarily in the first 4 weeks after calving (Grummer, 1993), and as a secondary disease of other production diseases that depress appetite or increase body fat mobilization. The clinical symptoms comprise depression, lack of appetite and weight loss, and the cows are weak and apathetic (Radostits et al., 2000). Most cows suffer from non-specific clinical signs including reduced rumen motility and decreased milk yield. However, this disease occurs especially in its subclinical form and can be a problem for up to 50% of cows in early lactation (Ingvartsen, 2006).

Risk factors for fatty liver in dairy cows may be nutritional, managerial, and genetic. Prepartum risk factors include obesity, severe feed restriction, feeding excess energy, and long calving interval. Postpartum risk factors are different diseases and infection, fasting and feed restriction, as well as ketogenic diets and sudden feed changes (Bobe et al., 2004). Among the highest risk factors of fatty liver is a high rate of mobilization of body lipids around calving in overconditioned cows which have been overfed in late lactation and the dry period and had low feed intake around calving (Ingvartsen, 2006). Increased prepartum body weight gain or body condition increase liver fat content and indicate increased risk of fatty liver and related disorders. Therefore, the primary nutritional risk factor for fatty liver is obesity. In obese cows, lipolysis of adipose tissue is increased more during peripartal period than in cows with normal body condition. Obese cows have a greater decrease in feed intake around calving and, therefore, have a more severe negative energy balance (Bobe et al., 2004). During the negative energy balance body fat is mobilized into the bloodstream in the form of non-esterified fatty acids (NEFA). NEFA are taken by the liver in proportion to their supply, but the liver does not have capacity sufficient to oxidize and use all amount of NEFA for energy. Therefore, cows are predisposed to accumulate NEFA as triglycerides within the liver when large amounts of NEFA are released from adipose tissue (Overton & Waldron, 2004). Rate of triglyceride synthesis is proportional to plasma NEFA concentration, thus fatty liver is likely to develop when plasma NEFA are elevated (Grummer, 1993).

Severe negative energy balance caused by low feed intake around calving leads to increase in body fat mobilization and plasma NEFA concentration. Thus, the severity of fatty liver may be reduced if drop in feed intake before calving is prevented (Bertics et al., 1992).

Diets deficient in cobalt (Co) have been shown to cause fatty liver in sheep. It has been termed as ovine white liver disease. Hepatic lipidosis in goats due to feeding low levels of Co in their diet has also been reported (Johnson et al., 2004).

Fatty liver is associated with decreased health status and reproductive performance. In severe cases, milk production and feed intake are also decreased (Bobe et al., 2004). Although fatty liver is often a reversible condition, it predisposes cows to reduced liver function and to a number of other production diseases (Ingvartsen, 2006). The incidence of fatty liver is strongly associated with the incidence of especially ketosis and displaced abomasum because these disorders are related to severe negative energy balance too (Bobe et al., 2004). Inhibition of gluconeogenesis may also occur when triglycerides accumulate in liver (Overton & Waldron, 2004) and additionally increase the risk for ketosis.

Different aspects of the immune response are also suppressed in cows with fatty liver (Bobe et al., 2004; Overton & Waldron, 2004) and those animals are more prone to infectious diseases. However, the exact costs of fatty liver are difficult to estimate because condition can be diagnosed only by liver biopsy (Bobe et al., 2004).

## 2.2. Ketosis

Ketosis (acetonemia) is a metabolic disorder characterized by elevated concentrations of the ketone bodies (acetoacetate, beta-hydroxybutyrate and acetone), and a low to normal concentration of glucose in the blood. This disorder occurs both subclinically and clinically. Prevalence of subclinical ketosis is the highest in the first 2 weeks of lactation ranging from 8,9% to 34% in various studies (Ingvartsen, 2006; Rushen et al., 2008). Blood glucose in clinically affected cows fall below the level required to support nerve and brain function and cows often exhibit signs of central nervous system dysfunction. Ketotic cows also suffer inappetence which further exacerbates the negative energy balance. Milk production falls seriously which helps the cow to cope with the negative energy balance (Goff, 2006a).

Four main types of ketosis are: primary ketosis, secondary ketosis, butyric acid ketosis and underfeeding ketosis (Ingvartsen, 2006). Classical or primary ketosis (also called spontaneous or production ketosis) generally occurs in cows during the first 2 to 4 weeks of lactation (Goff, 2006a). It develops when the glucose demand exceeds the gluconeogenesis capacity of the liver resulting in increased ketogenesis, and thus in high concentrations of ketone bodies in blood, milk, and urine. The disorder is mainly seen in obese cows. Secondary ketosis results from another disease that depresses feed intake and increases body fat mobilization (Ingvartsen, 2006).

Butyric acid ketosis is caused by large amounts of butyrate in the feed and probably by related depressed feed intake. Silage with high butyrate concentrations results in an increased concentration of beta-hydroxybutyrate in the blood, and such silage is consumed in lower amount then normal one. Underfeeding ketosis occurs especially in cows that are fed insufficiently. Underfed animals are deficient in glucogenic precursors and this condition then leads to increased ketogenesis (Ingvartsen, 2006).

Another form of ketosis often seen in US dairies usually occurs in cows less than 10 days in lactation. It can often be difficult to treat as is generally accompanied by some degree of fatty liver (ketosis/fatty liver complex). Ketosis, like fatty liver, also occurs during periods of elevated plasma NEFA. Clinical symptoms are similar to classical ketosis, but ketotic state was actually preceded by an increase in fat accumulation in the liver (Goff, 2006a).

Major factors directly or indirectly increasing the risk of fatty liver and ketosis are: overconditioning at calving, excessive mobilization of body fat, low nutrient intake, some nutrient or diet specific factors and management and environmental stress (Ingvartsen, 2006). Ketosis is a metabolic condition that occurs primarily when a cow is in negative energy balance immediately after calving. To support the energy demands, the body mobilizes fat reserves. Excessive release of NEFA from fat depots may overwhelm the capacity of the liver to use the fatty acids as a fuel. They are instead converted to ketone bodies (Rushen et al., 2008). Feed intake is naturally depressed by about 20% around the time of calving. Simultaneous with the feed intake depression, plasma NEFA and liver triglyceride concentrations exhibited their greatest increase (Goff, 2006a). All factors exacerbating negative energy balance and depression of feed intake around calving increase the risk of ketosis, especially in overconditioned cows.

Negative energy balance, high blood levels of NEFA or beta-hydroxybutirate and ketosis reduce the response capacity of white blood cells so that invading bacteria can out compete the immune system of the cow (Zadoks, 2006). Hypoglycemia alone is not likely to exacerbate periparturient immunosuppression, but hyperketonemia appears to have multiple negative effects on immune functions (Overton & Waldron, 2004). Impaired immune functions around calving make cow more prone to infectious diseases. Cows that have suffered ketosis have a higher risk of ketosis in the following lactation (Ingvartsen, 2006). Affected animals produce 0,7-3,3 kg/day less milk in the rest of lactation (Fourichon et al., 1999) or about 200 kg annually (Guard, 1996). In addition, prolonged and pronounced negative energy balance in early lactation adversely affects reproductive performance of cows and increases the risk of other metabolic diseases (Rushen et al., 2008).

Ewes experience ketosis typically during the last month of pregnancy and is known as pregnancy toxemia (Schlumbohm & Harmeyer, 2003). Ovine pregnancy toxemia (twin lamb disease, pregnancy disease) occurs primarily in ewes carrying more than one fetus (Andrews et al., 1996; Schlumbohm & Harmeyer, 2003). It is usually seen when the plane of nutrition has been static or falling over the last month of gestation when the requirements of the ewe are increasing to allow the growth of the fetuses (Andrews et al., 1996). In goats, two clinical form of ketosis have been described: pregnancy toxemia during the last month of pregnancy and primary ketosis during the first month of lactation (Stelletta et al., 2008).

Limitation of the availability of glucose has been found to be a crucial factor for the development of pregnancy toxemia. Hyperketonemia lowered plasma glucose concentration and depressed endogenous glucose production in sheep by approximately 30%, and this facilitates the onset of pregnancy toxemia (Schlumbohm & Harmeyer, 2003).

Mortality is high if pregnancy toxemia is not treated, with only about 20% of affected ewes recovering without treatment. Encephalopathy can results from depressed glucose metabolism in the brain (Andrews et al., 1996).

## 2.3. Rumen Acidosis

Ruminal acidosis is a nutritional disorder of ruminants generally resulting from ingestion of large amounts of feeds rich in readily fermentable carbohydrates, particularly when animals have not been gradually acclimated to those feeds (Bramley et al., 2008). The disorder usually occurs in high-producing dairy cows. Subacute or subclinical ruminal acidosis (SARA) is considered to be one of the major threats to the welfare of lactating dairy cows and may affect up to 20% of cattle in early to mid lactation (Rushen et al., 2008). Subclinical rumen acidosis is defined as a condition where rumen fluid pH is below 6,0 while acute rumen acidosis is when rumen pH is below 5,5 associated with rumen motility that is weak or ceased (Ingvartsen, 2006).

Rumen acidosis classically occurs when an animal consumes the excess of grain (Stone, 2004). Fermentation of the high grain diet reduces rumen pH which can cause undesirable changes in microbial populations within the rumen (Goff, 2006a). Rumen pH is lowered due to large quantities of volatile fatty acids and lactic acid produced during grain fermentation (Bramley et al., 2008). Lactic acid production is not the hallmark of rumen acidosis in dairy cows, as observed in beef feedlot cattle. Instead, it is the total organic acid load induced by the high grain diet combined with the inability of the cow to buffer the acids with salivary secretions (Goff, 2006a; Stone, 2004). However, lactate accumulation may occur in cows after calving if the shift in fermentable carbohydrates between the diets fed before and after calving is too dramatic (Stone, 2004).

Low rumen pH can cause rumenitis, metabolic acidosis, lameness, hepatic abscesses formation, pneumonia, and even death (Bramley et al., 2008). Reduced ruminal efficiency, liver and lung abscesses, and laminitis all can be related to SARA (Stone, 2004).

## 2.4. Laminitis

Laminitis and laminitis-related hoof problems (sole ulcer, white line abscess, solar hemorrhage, etc.) are one of the leading causes of lameness in cows. Laminitis has been associated with nutrition, and specifically with ruminal acidosis either in its acute or subacute form. Exact relationship between laminitis and SARA is not known. One of the theories states that damage of the ruminal epithelium induced by acidosis allows absorption of histamine and endotoxins into the blood. These and possibly other compounds affect circulation within the hoof and cause inflammation leading to the condition known as laminitis (Stone, 2004). Cows fed the higher level of crude protein may have increased incidence and duration of lameness. It is considered that products of degradation of protein excess in the rumen may be the causative agents for lameness.

It is widely accepted that nutrition helps maintain claw health through the production of good quality horn (Baird & Muelling, 2009). For example, heifers fed on hay before calving might have better foot health than those fed silage, even if both groups are being fed the same diets postpartum. Forage type and level of moisture in the diet could influence lameness and this effect is exerted even before first calving (Logue & Offer, 2001).

It has also been shown that nutritional supplements such as biotin and zinc (Zn) can help reduce lameness through improving claw horn quality (Baird & Muelling, 2009; Goff, 2006a). Biotin is essential for two major metabolic pathways in keratinisation, keratin protein synthesis and lipogenesis. Other vitamins also may play important roles in maintaining claw integrity, including vitamin A and vitamin E. Zinc, as a component of many enzyme systems, has a role in major functions during keratinisation including the formation of structural proteins (Baird & Muelling, 2009). This element improves claw integrity by speeding wound healing and epithelial tissue repair. Other trace minerals that impact claw condition include iodine (I), selenium (Se), copper (Cu), manganese (Mn) and Co. Calcium and phosphorus (P) are also needed for normal claw growth and integrity.

Laminitis reduces profitability of the dairy herd. It is estimated that 15% of cows culled for slaughter are culled due to laminitis. In clinically lame cows, milk yield was reduced from 4 months before and for the 5 months after treatment. The total mean estimated reduction in milk yield per lactation was approximately 360 kg (Green et al., 2002).

## 2.5. Displaced Abomasum

Displaced abomasum is a multifactorial disease where the abomasum is dilated as a result of gas accumulation and dislocated to the left (left-displaced abomasum) or to the right (right- displaced abomasum) in the stomach in relation to the normal placing (Ingvartsen, 2006). The passage of feed to the intestines is partly or totally blocked. Approximately 80-90% of the incidences are left-displaced abomasums. The disease is most frequent in high producing cows in early lactation and 80–90% of the cases are seen in the first 4 weeks postpartum (Shaver, 1997). It occurs in approximately 3,5% of dairy cows each year (Goff, 2006a).

Nutrition has been implicated as a major risk factor in the etiology of displaced abomasums, but the precise cause is still unclear. Low feed intake and increased negative energy balance prepartum have been found to increase the risk of displaced abomasums in cows (Goff, 2006a; Ingvartsen, 2006). Cows fed high concentrate diets in early lactation or diets with inadequate particle size are also at increased risk of displaced abomasums (Goff, 2006a). Sudden changes in the diet and rapid increase in concentrate allowance in early lactation are also the risk factors. Some feeds increase the risk of the disease compared to others. Thus, a higher risk of displaced abomasum has been found for cows fed silage compared to those fed hay, probably because silage is often finely chopped. Risk can be almost eliminated if cow eats a kilogram of straw daily (Ingvartsen, 2006).

Displaced abomasum develops if three major conditions are met. The first one is reduced contractility and atony of abomasum which is gas-dilated. The next condition is that the mesentery shall stretch for the abomasum to be able to dislocate, and a third condition is the space in the abdominal cavity (Ingvartsen, 2006). The conditions leading to atony and reduced motility are still unclear, but hypocalcaemia around calving is a possible factor. Reduction in blood Ca concentration around calving results in a reduction of abomasal contractility, which can lead to its atony and dilatation (Shaver, 1997). Hypocalcemic cows after calving have 3,4 to 4,8 times increased risk of development of abomasal displacement (Massey et al., 1993).

Volatile fatty acids (VFA) in the abomasum also have been reported to reduce abomasal motility. Effects of VFA on motility may be exacerbated by low ruminal absorption of VFA during the transition period (Shaver, 1997). Hypocalcaemia also may play a role in this process. Due to reduced feed consumption, slow contractions and inadequate filling of the rumen which therefore does not reach the ventral abdominal wall, empty space appears for movement of abomasum. Then usually more fatty acids escape absorption in the rumen and reach the abomasum. Those VFA, along with hypocalcaemia and accumulated gases, contribute to reducing abomasum contractility and development of atony (Goff & Horst, 1997c). Thus, inadequate feed consumption and insufficient rumen fill with reduced motility and strength of abomasal contractions together contribute to the onset of this disorder (Goff & Horst, 1997c).

Low intake of concentrates during the prepartum period also may increase the risk of left displaced abomasum because absorptive capacity of the ruminal papillae is not increased sufficiently and microbial population of the rumen is not adapted prior to intake of high energy postpartum diets (Shaver, 1997). Too rapid increase of concentrates after calving may reduce roughage intake and potentially increase the risk of displaced abomasum (Ingvartsen, 2006). Uncomplicated ketosis, fatty liver, retained placenta, and metritis are also risk factors for left displaced abomasums (Bobe et al., 2004; Shaver, 1997).

Treatment of displaced abomasum often requires surgical treatment, and deaths are not uncommon so that potential losses due to disease are very high. In addition, cows that recover from abomasal displacement produce about 350 kg less milk during the next month (Shaver, 1997) or 0,8-2,5 kg/day in the rest of lactation (Fourichon et al., 1999). According to Guard (1996) the average losses amounted to about 380 kg of milk annually.

## 2.6. Milk Fat Depression

Nutrition influences both the quantity and composition of milk fat. In modern dairy production cows are fed diets with high level of concentrates to maximize milk production but such diets often causes drop in milk fat. Condition is known as milk fat depression (MFD) or low-fat milk syndrome. Although milk volume and yield of other milk constituents may not be affected or may be even increased, depression of fat in milk can be a serious economic problem for dairy producers. This syndrome is not a disease, but rather metabolic consequence of attempts to reach the higher milk production in animals.

Milk fat depression refers to a marked reduction in milk fat yield with no change in milk yield or yield of other milk components (Harvatine et al., 2009). It represents a level of milk fat production below the genetic potential of the cow, usually as below 3,2% fat in milk for Holstein, or below 4,2% fat in the milk for a Jersey herd. When problem with MFD exists, the ratio of milk fat/milk protein will be less than 1,0 for Holstein herds. In severe cases the ratio will be less than 0,8. Milk fat yield can be reduced up to 50% and the effect is specific for milk fat (Bauman et al., 2008). The problem has been observed in many feeding situations and dietary conditions, including high level of concentrates and low level of dietary fiber, and diets supplemented with unsaturated oils. The fat content of milk can also be affected by the physical characteristics of the roughage, including grinding and pelleting (Perfield & Bauman, 2005). In diet-induced MFD, the yield of all individual fatty acids is reduced, but the decline is greatest for short- and medium-chain fatty acids that are synthesized de novo in mammary gland (Bauman et al., 2008).

It is generally accepted that induction of MFD requires both an altered rumen environment and the presence of unsaturated fat in the rumen which biohydrogenation pathways are altered (Perfield & Bauman, 2005). Changes in ruminal microbial processes are an essential component for the development of MFD and are often associated with a decrease in rumen pH and a shift in the acetate/propionate ratio.

Historically, there have been several theories proposed to explain the basis for diet-induced MFD. The biohydrogenation theory was proposed by Bauman & Griinari (2001) focusing on the latest research in MFD problem. The authors sugested that "under certain dietary conditions the pathways of rumen biohydrogenation are altered to produce unique fatty acid intermediates which are potent inhibitors of milk fat synthesis". The theory was born after identification of trans-10,cis-12 conjugated linoleic acid (CLA) as a highly potent inhibitor of milk fat synthesis (Bauman &

Griinari, 2001). This isomer is a rumen precursor of trans-10 18:1 in biohydrogenation pathway that may arise under certain dietary conditions, including high concentrate/low fiber diets. The biohydrogenation of linoleic acid in rumen involves the formation of trans-11 18:1 and cis-9,trans-11 CLA as the main intermediates. However, under certain dietary situations the other pathways of biohydrogenation can happen and some of the intermediates produced during the process may be potent inhibitors of milk fat synthesis (Bauman & Griinari, 2001). The most extensively studied is trans-10,cis-12 CLA. Minimal quantities of this isomer markedly inhibited milk fat synthesis and a curvilinear reduction in milk fat yield occurred with increasing trans-10,cis-12 CLA in the rumen content (Baumgard et al., 2001).

Isomer trans-10,cis-12 CLA alone does not completely explain the extent of the decrease in milk fat. Additional biohydrogenation intermediates produced in the rumen probably inhibit milk fat synthesis and two of them (trans-9,cis-11 and cis-10,trans-12 CLA) have already been identified (Bauman et al., 2008).

Consequences of MFD are almost exclusively related to losses in yield of milk fat and lower price of milk. As problem is normally reversible, health consequences are usually not a matter of concern. However, MFD inducing diet is usually risky for rumen homeostasis due to high level of carbohydrates and low level of effective fiber which may reduce ruminal pH and cause acidosis (Bergen, 2009). If acidosis is not compensated and drop in ruminal pH is not prevented, the condition may become an important animal health issue.

## 2.7. Parturient Hypocalcaemia

The lack of Ca in the diet does not lead to any changes in health or production for a long time because it is compensated by mobilization of Ca reserves in the skeleton. The change of Ca content in the blood may not occur until the cows show symptoms of osteoporosis and bone fractures. A unique example, however, is an acute hypocalcaemia in dairy cows that occurs in the periparturient period as a result of sudden loss of great amounts of Ca in colostrum along with temporary dysfunction of the mechanisms of Ca mobilization from bones. From a relative inactivity in the dry period, these mechanisms often fail to achieve full activity quickly enough to maintain normocalcaemia around calving (Horst et al., 1994). More pronounced hypocalcaemia causes progressive neuromuscular dysfunction which is manifested as a specific clinical syndrome known as paresis puerperalis (milk fever, parturient paresis, calving paralysis). More often, however, and in a larger number of cows, hypocalcaemia exists in a subclinical form with very few or no symptoms (subclinical hypocalcaemia).

Hypocalcaemia was defined as the content of total Ca in blood below 2 mmol/L with or without clinical signs of paresis (Oetzel et al., 1988; Radostits et al., 2000), which is roughly equivalent to 1 mmol/L of ionized Ca (Massey et al., 1993; Oetzel, 1996).

Milk fever (MF) is non-febrile disease of adult dairy cows accompanied by general weakness, circulatory collapse and depression of sensation (Oetzel, 1988). It is one of the most common metabolic disorders in intensive dairy production (Horst, 1986) and a typical nutritional disorder (Ender et al., 1971). The disease attacks 5-10% of adult dairy cows in the U.S. and Europe annually, primarily those with high productivity. The classic syndrome of MF occurs immediately before or after parturition with about 87% of cases in the first 48 hours after calving and only about 9% cases before or during the parturition (Oetzel, 1988). The disease is characterized by pronounced hypocalcaemia which is accompanied by hypophosphataemia in most cases, and by more or less serious hypermagnesaemia (Phillippo et al., 1994).

Phenomena like paresis may rarely occur in beef cows, ewes (before or after lambing) and more frequently in dairy goats (Oetzel, 1988). In mares, the disease can occur several weeks after birth and is known as milk tetany or eclampsia (Radostits et al., 2000).

Milk fever has been recognized as a nutritional disorder with various endocrine abnormalities which occur more or less as a secondary phenomenon (Ender et al., 1971). The characteristic of the disease is an acute inability to mobilize Ca from bone (Horst, 1986) despite apparently normal endocrine dynamics in most affected cows (Radostits et al., 2000). The problem seems to lie in the reduced sensitivity of target tissues (bone, kidneys and intestines) on calcitropic hormones due to reduced number of receptors or because of their dysfunction (Horst et al., 1994).

Widely accepted theory for a long time was that MF was caused by a high content of Ca in the diet consumed before calving, as well as by unfavorable ratio of Ca/P. However, further studies emphasized importance of the high content of monovalent cations in the diet, mainly potassium (Block, 1984; Ender et al., 1971; Goff & Horst, 1997b; Horst

& Goff, 1997). As a relative excess of cations in the diet affects blood acid-base balance toward alkalosis, the disruption of integrity of tissue receptors is related to metabolic alkalosis that normally exists in cows at the time of calving (Goff, 2000). Thus, the primary cause of MF is considered to be the temporary inability of cow tissues to adequately respond to stimuli generated by calcitropic hormones, mainly parathyroid hormone (PTH) (Goff et al., 1991; Goff,2000).

In addition to monovalent ions, risk factors also include dietary magnesium (Sansom et al., 1983), dietary P (Curtis et al., 1984), and possibly vitamin D (Goff, 2000; Horst, 1986). Animal factors as the risks for hypocalcaemia include age (Curtis et al., 1984; Dishington, 1974) and breed (Goff, 2000;Oetzel, 1988). Adult cows are at increased risk for the disease. Occurrences are more frequent from the third lactation onwards, reaching the plateau of 12-15% in cows 6-10 years of age (Curtis et al., 1984; Dishington, 1974). The most predisposed breeds for milk fever are Jersey and Guernsey, followed by Holstein and Brown Swiss. In the case of Jerseys possible reasons are considered to be the higher milk production per unit of body weight (Oetzel, 1988) and a higher content of Ca in the colostrum (Goff, 2000). In addition, the number of receptors in the intestine is about 15% lower in Jerseys than in Holsteins (Goff, 2000; Horst et al., 1997).

Depending on the age, breed, feeding regimen and housing conditions, as many as 50-80% of cows can suffer MF annually in some herds (Oetzel, 1988). Concerning that well-timed treatment of MF easily solves the problem and is relatively inexpensive, the disease was not considered as a factor of economic importance in dairy production for a long time (Radostits et al., 2000). However, later findings confirmed its close relationship with other health disorders in the puerperium (Curtis et al., 1985; Massey et al., 1993). Many of these disorders are now considered as complications of hypocalcaemia. They generally occur as a complex, rarely as isolated diseases (Curtis et al., 1985) and include most of metabolic and reproductive disorders in periparturient period: dystocia, retained placenta, metritis, uterine prolapse, ketosis, mastitis and displaced abomasum (Horst et al., 1997). Risk of occurrence of each of them is much higher in hypocalcemic cows (Curtis et al., 1985). A well known physiological effect of hypocalcaemia is atony of smooth and skeletal muscles because their contractility is a function of Ca concentration in extracellular fluid (Ramberg et al., 1984). Most of the possible complications are related specifically to reduced contractility and to atony of smooth muscles of digestive and reproductive tract (Beede, 1995).

Spontaneous recovery of MF is usually not possible and approximately 75% of affected cows die if treatment fails (Oetzel, 1988). Even under normal conditions about 8% of treated cows still die due to various complications, 12% of them are culled (Guard, 1996) and about 25% require another treatment. Due to increased susceptibility to other health disorders and possible complications, the production life of cows that experienced MF was reduced by an average of 3-4 years (Horst et al., 1997).

Subclinical hypocalcaemia influences in the same manner as the clinic one, but to a lesser extent. As is more common in the herd, adverse effects of subclinical hypocalcaemia on herd economy can be equal or even greater than the effects of MF due to its broader influence on feed intake, secondary disease conditions, and milk production during early lactation (Horst et al., 1994).

## 2.8. Hypomagnesaemia

Magnesium is an essential mineral with many functions in the body but its homeostasis is not hormonally regulated. The concentration of Mg in the blood is dependent exclusively on its absorption from the diet. If Mg secretion in milk and its endogenous losses exceeds absorption from the forestomachs hypomagnesaemia occurs because of the lack of hormonal control. Hypomagnesaemia is common problem in ruminants and may be one of the major health problems of cattle, sheep and goats in large scale production systems especially in temperate climates (Mayland, 1988; Stelletta et al., 2008).

Clinical hypomagnesaemia is also called hypomagnesemic tetany, grass tetany, spring tetany, or lactational tetany. Clinical signs usually occur when the animal is both hypomagnesaemic and hypocalcaemic (Robinson et al., 1989). In its subclinical form, however, hypomagnesaemia may be risk factor for other diseases. Around calving it may be the important risk factor for MF (Sansom et al., 1983; Schonewille et al., 2008).

Hypomagnesaemia may develop due to Mg deficiency in the diet or due to its low utilization in the gut. It may also occur because of a need for increased amounts of Mg during parturition and early lactation (Mayland, 1988). Clinical hypomagnesaemia in cows with plasma Mg concentrations bellow 0,4 mmol/L is manifested as grass tetany (Schonewille et al., 2008). Grass tetany has been investigated extensively, but its complex etiology is not well understood. Large number of factors influences the development of the disease (Robinson et al., 1989). It appears within 2-4 weeks after cattle or sheep have been turned out to rapidly growing pasture. Ewes with twins are more susceptible to grass tetany than are ewes with single lamb (Mayland, 1988). Specifically, older lactating animals consuming lush, intensively fertilized, cool-season grasses during the early spring are the most frequently affected (Robinson et al., 1989). Older animals also have reduced ability to mobilize body reserves of Mg. Lower average environmental temperatures (<14°C) facilitate the onset of the disorder (Mayland, 1988). However, periods of rapid plant growth during any season, resulting in forage of low Mg and high moisture, high nitrogen (N) and K, present dietary conditions that increase the potential for the development of tetany (Robinson et al., 1989). Grass tetany also may develop if animals graze forage that is high in digestible protein, but low in digestible energy (Mayland, 1988).

Grass tetany generally occurs when the dietary intake of total Mg is not particularly low, but factors which increase the animal's requirement for Mg or reduce the availability of dietary Mg are present (Mayland, 1988). The Mg concentrations in forage and subsequently in the blood of cattle are influenced strongly by high amounts of fertilizer K and, to some extent, fertilizer N. Absorption of Mg by plants is reduced by high levels of K in the soil (Mayland, 1988; Robinson et al., 1989). Mg absorption by ruminants is also reduced by a high intake of K (Schonewille et al., 2008), and by high content of Ca and P in the diet (Sansom et al., 1983). High concentrations of N in the forage may additionally decrease the availability of Mg (Mayland, 1988).

Absorption of Mg is dependent on the concentration of Mg in the rumen fluid and the functionality of the Mg active transport process across the rumen wall. The active transport mechanism is critical for the animal when dietary Mg concentration is less than 0,25%. Several factors, such as dietary K, can inhibit Mg absorption by this pathway. At high concentration of Mg in the rumen it will be absorbed by passive transport, and this mechanism is not affected by high K but only by the solubility of Mg in the rumen fluid. This mechanism requires dietary Mg to be not less than 0,35% (Goff, 2006b).

Mg deficiency results in reduced appetite which decreases total nutrient intake (Robinson et al., 1989), thus chronic hypomagnesaemia results in reduced feed intake and milk production. It may progress quickly into acute hypomagnesaemia which terminates in convulsions, coma and death. One-third of animals with clinical symptoms die (Mayland, 1988).

Even mild form of hypomagnesaemia (Mg <0,85 mmol/L) significantly reduces the mobilization of Ca from the skeleton around calving (Goff, 2000;Sansom et al., 1983). That is why hypomagnesaemia may be a risk factor for parturient hypocalcaemia and milk fever. According to Sansom et al. (1983) possible reasons for inhibition of Ca mobilization are considered to be:

a) influence on PTH secretion; there are indications that serious Mg deficiency may impair or completely interrupt secretion of PTH,

b) interference with the action of PTH on target tissues; the integrity of interaction between PTH and its receptors is disrupted and activity of initiated enzymes is disabled,

c) interference with the metabolism of vitamin D; the first phase of vitamin D hydroxylation in the liver requires Mg2+ ions, while the second phase in the kidney takes place in the presence of PTH.

## 2.9. Udder Edema

Udder edema occurs sporadically in cows and heifers near parturition, peaking in severity during the immediate prepartum period. Unless complicated, recovery of edema is spontaneous within a few days after calving. However, if edema is sufficiently severe it can interfere with suckling by the calf, milking, and may cause other complications including udder inflammation. Edema is caused by excessive accumulation of fluid underneath the skin and at least some mammary edema is associated with pregnancy and parturition, especially in primigravid heifers (Malven et al., 1983).

Investigations of the causes of udder edema and the development of strategies to reduce its prevalence have progressed in very limited degree during last several decades (Goff, 2006a). Some authors (Schmidt & Schultz, 1959) proposed that severity of udder edema is based on an inherent physiological phenomenon, because a cow may tends to have the same amount of edema each year regardless of the feeding. Feeding program had little if any effect on udder edema (Randalu et al., 1974). "Steaming up" method of feeding dry pregnant cows and high level of concentrates in their diet was initially blamed for high incidences of udder edema. However, investigation found that the amount of edema at calving was not related to grain feeding during the dry period or after calving (Schmidt & Schultz, 1959). Edema also is not correlated with the body condition of the cows (Schmidt & Schultz, 1959). However, addition of Na or K to the diet before calving can increase the incidence and severity of udder edema in dairy cows and first calf heifers (Nestor et al., 1988). In practical terms, high level of NaCl in the feed during the dry period may be one of the major factors in the development of udder edema around calving. Conflicting results from some earlier studies on effect of grain on udder edema could be explained with salt content of the diets. However, KCl as a replacement for NaCl results in edema of about the same severity (Randalu et al., 1974).

Cows with severe edema that requires veterinary treatment are more likely to have additional health problems. Authors found higher culling rates for cows with more severe edema. Thus, there would be a substantial economic benefit to minimizing edema in dairy herds (Dentine & McDaniel, 1984).

## 2.10. Retained Placenta

Inadequate antioxidant status or "oxidative stress" of the cow contributes to a poorly functioning immune system and increases the risk of mastitis as well as retained placenta (retained fetal membranes, placental retention). Fetal membranes are retained if not expulsed longer than 12 hours after parturition. Se and vitamin E are important dietary antioxidants and their low levels in the diet are associated with a high incidence of mastitis and retained fetal membranes (Goff, 2006a). Subsequently, addition of vitamin E or Se may improve antioxidant status and decrease the incidence of those diseases. The similar effects can be obtained by addition of beta-carotene in the diet. Blood lymphocyte proliferation was higher in cows supplemented with beta-carotene, and phagocytic activity of blood neutrophils was enhanced as well as intracellular killing by blood neutrophils. Therefore, dietary beta-carotene can elevate blood beta-carotene and enhance peripartum host defense mechanisms by enhancing lymphocyte and phagocyte function (Michal et al., 1994).

Hypocalcaemia, among others, may be an important risk factor in the development of retained placenta. Muscle weakening or absence of uterine contractions in hypocalcemic animals does not contribute to the expulsion of fetal membranes (Goff & Horst, 1997c; Oetzel, 1988). In cows with hypocalcaemia placental retention is 3,2 to 4 times more frequent than in normocalcemic cows (Curtis et al., 1985). Hypocalcaemia also delay the physiological involution of uterus and increases the incidence of metritis (Beede, 1995). Moreover, hypocalcaemia is considered as one of the main causes of uterine prolapse, also due to loss of uterine muscles tone.

Cows with retained placenta were 3 times more likely to develop mastitis than animals without retained placenta (Overton & Waldron, 2004). It has been reported that placental retention, metritis and mastitis predisposes dairy cows to foot problems. Moreover, uncomplicated ketosis, retained placenta, metritis, and hypocalcaemia at parturition are also risk factors for left displaced abomasums (Shaver, 1997). Cows with retained placenta or metritis produce 0,3-2,3 kg/day less milk during subsequent lactation (Fourichon et al., 1999). In the case of retention alone, Guard (1996) stated that losses are on average 350 kg of milk annually.

## 2.11. Metritis

Metritis and endometritis are inflammatory uterine diseases. They frequently occur soon after calving and may severely compromise reproductive performances. Metritis and endometritis refer to the inflammation of the uterus and of its endometrial lining - both conditions are referred to subsequently as metritis (Urton et al., 2005). Younger cows were more likely to have dystocia or assisted deliveries, while older cows were most likely to have retained placenta and metritis (Lewis, 1997).

In some herds, up to 40% of the postpartum cows may be diagnosed with, and treated for uterine infections. However, the exact causes of uterine infections are unknown but are associated with several factors. Cows with dystocia, retained placenta, or stillbirths, and other metabolic disorders are more likely to develop metritis than healthy cows. Impaired immune functions before and after calving seem to predispose cows to severe uterine infections. It has been suggested that the function of neutrophils is impaired in cows that develop uterine infections. Thus, methods for regulating immune function in periparturient cows may have potential for preventing uterine infections. However, prevention of uterine infections is still difficult because the primary causes cannot be defined clearly (Lewis, 1997). Malnutrition influences the ability of the immune system to function, which affects the incidence of diseases such as mastitis and metritis (Goff, 2006a).

Cows suffering metritis exhibit reduced milk yield and reproductive performances (Urton et al., 2005). Few cows die from uterine infections, but affected animals are more likely to be culled for poor reproductive performances. The estimated cost to producers for each cow with metritis was 106 U.S. dollars (Lewis, 1997). Many of the financial costs of metritis are indirect, such as increased days open or predisposition to other diseases, and are thus difficult to measure. As mentioned previously, metritis is one of the diseases that predispose dairy cows to foot problems and left displaced abomasums (Shaver, 1997).

In contrast, the effects of metritis on milk production can be measured immediately, and losses during first four months after calving can be almost 270 kg. Also, besides reduced milk production in sick animals, some pharmaceuticals for treatment of uterine infections contaminate milk with residues, and the milk must be discarded. This is how uterine infections may have an indirect effect on milk production (Lewis, 1997).

# 3. USE OF BLOOD AND MILK ANALYSES TO EVALUATE NUTRITIONAL AND DISEASE STATUS

If recognition of subclinical diseases is difficult, the condition may be confirmed by analyzing blood, milk or sometimes urine, although some of them are still difficult to diagnose in practice, including rumen acidosis and fatty liver (Ingvartsen, 2006). Blood and milk analyses as tools to evaluate nutritional and disease status of individual animal or whole herd are aimed to help making decisions for improvement of nutritional strategies and production management. However, many problems are associated with interpretation of results of laboratory analyses, including Compton metabolic profile test, as well as other tools for assessment of metabolic status of dairy animals.

## 3.1. Milk Analyses

Producers can use changes in milk production or in milk composition to monitor the health of their animals, but these tools are not always completely reliable because milk production corresponds poorly with mild or subclinical infectious disease. Nevertheless, measuring energy balance from changes in milk composition, most likely changes in milk fat and protein contents, could provide a cheap and reliable estimator of energy balance. A study clearly showed that there was a strong relationship between energy balance and milk composition under stable feeding conditions (Friggens et al., 2007). Milk composition varies with energy status and was proposed for measuring energy balance on-farm. Ratio milk fat/protein below 1,4 indicates the optimal or positive energy balance, and above it balance is negative (Pehrson, 1996; Zadoks, 2006). During peak lactation many cows with negative energy balance have this ratio in milk even above 2,1 (Pehrson, 1996).

The biological basis for the relationship between fat/protein ratio and energy status of the animal can be found in two physiological features: a) mechanism that maintains milk energy output by increasing milk fat content when yield is compromised due to a deficit in energy supply, and b) decreased milk protein content under negative energy balance (Friggens et al., 2007). Negative energy balance also may be reflected in elevated ketone bodies which are excreted in milk and urine. Milk has about half the ketone level of blood, and is recommended to check milk ketone levels for detection of ketosis.

Milk fat depression is an example of disorder that can be diagnosed exclusively by milk analysis. Specific blood test for MFD diagnosis does not exists, but milk analyses are usually enough to evaluate risk for or presence of MFD in a herd. Every drop in milk fat percentages with no changes in the content of other milk constituents and milk yield can be considered as MFD. In serious cases, diet-induced MFD can result in reduction of milk fat yield of up to 50% or even more (Bauman & Griinari, 2001; Bauman et al., 2008) and represents one of the greatest risks for production economy.

## 3.2. Blood Analyses

Analyses of blood constituents have long been a matter of concern as a possible tool for assessment of diet adequacy or metabolic and health status of animals. Compton metabolic profile test (CMPT) originally involved the analysis of a set of blood variables from three groups of seven cows, one near peak lactation, and others in midlactation and in late dry period (Kronfeld et al., 1982). The samples are collected at least three times yearly: summer, autumn and winter, or when nutritional imbalance is expected, using the same procedures and timings (Radostits et al., 2000). Means of variables are calculated for each lactational group in each herd, and this set of means constitutes the metabolic profile. Each group mean is compared to reference ranges determined from corresponding means plus or minus two standard deviations that are based on group-means for all herds (Kronfeld et al., 1982). The test was based on the concept that laboratory measurement of certain components of the blood will reflect the nutritional status of an animal with or without presence of clinical abnormalities. However, the results of research indicate that CMPT may be useful only as an aid in the diagnosis of the nutritional imbalances and production diseases. It must be carefully planed and is still expensive. Laboratories with automated analytical equipment should be available and this is often a major limiting factor (Radostits et al., 2000).

Compton metabolic profile test includes following analyses: blood glucose, packed cell volume, hemoglobin, blood urea nitrogen, serum inorganic phosphate, serum Ca, Mg, K and Na, total serum protein, albumin and globulins, serum Cu and iron (Fe), and plasma NEFA. Obtained results are interpreted with other relevant information taken on the day of sampling related to individual animals or to the herd: age, milk yield, days in milk, concentrates and forage intake, etc (Radostits et al., 2000).

Workers tested the CMPT as a guide to the nutritional status of dairy cows, and found that blood variables were not reliable predictors of energy and nutrients consumption relative to the requirements. Then they supposed that, for prediction of nutritional status, selection of blood variables should be probably different. Those workers have suggested, however, that protein intake may be reflected in blood concentrations of urea, albumin, and hemoglobin. Similarly, blood glucose concentration has attracted attention as an index of energy intake, although results from study to study have been inconsistent (Kronfeld et al., 1982).

Despite limitations and unconformity of metabolic profile tests, blood analyses still can help in diagnostics of some nutritional imbalances. Negative energy balance in postpartum cows can be detected by change in concentration of some blood metabolites (Rushen et al., 2008). As condition that affect immune functions and predispose animals to many metabolic disorders, it is important that negative energy balance is reflected in a higher levels of NEFA and beta-hydroxybutyric acid (BHBA) in the blood. To screen a herd for negative energy balance, Zadoks (2006) recommends having at least 12 animals tested. Testing of NEFA is done 2-14 days before calving and BHBA testing at 2-21 days after calving. If more than 10-15% of animals have NEFA levels above 0,40 mEq/L or BHBA levels above 1,36 mmol/L, the herd is considered to be suffering from excessive negative energy balance. Ketosis and fatty liver are diseases typically associated with negative energy balance for long periods of time. Increased ketone bodies and NEFA in the blood, and decreased blood glucose level are common findings in cows with ketosis and fatty liver. Values of glucose below 2,2 mmol/L are considered subnormal. However, fatty liver in cows currently can be diagnosed only by liver biopsy (Bobe et al., 2004).

The concentration of BHBA in the blood may be taken as a guide for correction of ration supplement for ewes during the final weeks of pregnancy to prevent pregnancy toxemia. It has been recommended to check 10% of the flock and feed is increased if blood concentration of BHBA exceeds 0,8 mmol/L (Morgante, 2004). Besides hypoglycemia and hyperketonaemia, plasma cortisol levels also may be elevated in sheep with pregnancy toxemia (Andrews et al., 1996). Early in the disease, both does and ewes will show a positive test for ketone bodies in the urine (Morgante, 2004).

There are no specific blood variables for detecting metabolic diseases like laminitis, displaced abomasums, udder edema, retained placenta and metritis. However, if these diseases are herd problems, blood metabolic profile should be checked for more than one variable, and carefully used as an aid for development of preventive strategies. When laminitis is a problem, a metabolic profile on both dry and lactating cows as well as springing heifers should include both red and white blood cells count, packed cell volume, Se, Zn, Cu, Fe, blood urea nitrogen, vitamins A and E, and beta-carotene. When retained placenta is a herd problem, blood profile should include serum minerals, Se, vitamin E, and beta-carotene. In individual cases, blood urea nitrogen and packed cell volume should be included. Metritis requires testing metabolic profile on dry and fresh cows which should include white blood cells count, Se, Zn, Cu, Fe, Mg, blood urea nitrogen, vitamins A and E, and beta-carotene.

Use of blood analyses to evaluate degree of hypocalcaemia is sufficient diagnostic test for this disorder. Concentration of Ca may be determined in whole blood, but more often determination is performed in plasma or serum. When there are no disturbances in acid-base status and protein metabolism, ionized and total Ca are strongly correlated and therefore total Ca concentrations in plasma may serve as an acceptable diagnostic value (Oetzel, 1988).

Hypocalcaemia means that the content of total Ca in the blood is below 2 mmol/L, or below 1 mmol/L of ionized Ca (Massey et al., 1993; Oetzel, 1996). Signs of paresis can occur at total Ca level of 1,9 mmol/L, but most of the cows remain on their feet down to 1,0-1,25 mmol/L. A reduction in the plasma Ca concentration at parturition is usually accompanied by hypophosphataemia and hypermagnesaemia (Phillippo et al., 1994). However, this is not always the case, especially if hypocalcaemia is not severe and is not clinically manifested (Crnkic et al., 2010; Joyce et al., 1997;Oetzel et al., 1988).

Blood analysis is also sufficient diagnostic test for hypomagnesaemia. Based on blood serum or plasma Mg levels, animals may be normally magnesemic (0,74-1,23 mmol/L), chronically hypomagnesemic (0,20-0,74 mmol/L) or acutely hypomagnesemic (<0,20 mmol/L) (Mayland, 1988). Sampling and analysis of the blood of several cows within 12 hours after calving is a good indicator of Mg status of the periparturient period. If serum Mg concentrations are less than 0,82 mmol/L in 90% of cows it suggests insufficient dietary Mg absorption (Goff, 2006a).

# 4. NUTRITIONAL STRATEGIES TO REDUCE METABOLIC DISEASES INCIDENCES

Although metabolic disorders are not easily categorized as to their cause, nutritional strategies have been developed to help prevent many of these disorders (Goff, 2006a). Scientific justification of those measures can be found in many research articles published in scientific journals during the last decades. However, problems in application of the strategies in the field still exist, as well as their limitations and possible negative consequences on milk production economy in certain circumstances. Strategies are based on major factors directly or indirectly increasing the risk of diseases such as overconditioning at calving, excessive mobilization of body fat, low nutrient intake, nutrient or diet specific factors and management and environmental stress (Ingvartsen, 2006). Strategies are expected to be able to reduce morbidity and at the same time improve reproduction and production.

To prevent metabolic disorders in the periparturient period nutritional strategies must start prior to calving. In a survey conducted by Curtis et al. (1985) consumption of nutrients before calving was directly related to the occurrence of metabolic disorders, and directly or indirectly to the occurrence of reproductive disorders after parturition. In most cases, these disorders occur as a complex and many of them are interrelated in their occurrence (Beede, 1995; Curtis et al., 1985). Consequently, strategies to reduce one disease can help preventing others. For example, strategies to reduce liver triglyceride accumulation at calving may decrease incidence of ketosis, etc.

Goff & Horst (1997c) recommended that three basic physiological functions must be maintained during the periparturient period if disease is to be avoided: adaptation of the rumen to lactation diets that are high in energy density, maintenance of a strong immune system, and maintenance of normocalcaemia. Whenever one or more of these functions are impaired, the incidence of both metabolic and infectious diseases is increased. However, measures to maintain these physiological functions must be conducted carefully avoiding other risk factors such as overconditionig and excessive mobilization of body fat around calving.

Metabolic disorders that may appear in the rest of lactation also require adequate prevention strategy to be controlled, including rumen acidosis, laminitis, and hypomagnesaemia and milk fat depression.

## 4.1. Adaptation Of The Rumen

Fully adapting the rumen flora to a high starch diet that will be fed after parturition requires about 3 to 4 weeks, and full development of rumen papillae requires about 5 weeks of concentrate feeding (Goff & Horst, 1997c). It is, therefore, important to start increasing concentrates in the diet 3-4 weeks before calving and continue during the first 1-2 weeks after parturition to fully adapt rumen to lactation diet. If fresh cow is abruptly switched to a high starch lactation diet, the risk of developing rumen acidosis exists because the lactate production and accumulation. During the rest of lactation, the most reliable means of preventing rumen acidosis is to apply feeding methods that ensure a more even distribution of feed intake over the day. Feeding TMR only once or twice a day may result in that cows eat an excessive amount of feed during a short period of time, because cows are strongly attracted by the arrival of fresh food (Rushen et al., 2008). Buffering agents such as Na-bicarbonate or alkalinizing agents such as Mg-oxide are added to high concentrate ration to reduce the risk of acidosis (Goff, 2006a). Na-bicarbonate should be supplemented particularly to corn silage-based diets at the rate of approximately 0,8 to 1% of DM (Stone, 2004). Too rapid increase in concentrate allowance during early lactation may reduce roughage intake and increase the risk of not only rumen acidosis, but also displaced abomasums (Ingvartsen, 2006).

A highly significant relationship between forage neutral detergent fiber (NDF) content in the diet and ruminal pH has been found. NRC (2001) recommended 19% of forage NDF as absolute minimum when formulating rations in the field. A system of "physically effective" NDF (peNDF) relates the ability of a feedstuff to stimulate chewing relative to a hypothetical long grass hay containing 100% NDF. The peNDF of a feed is the product of its physically effective fiber (pef) and NDF content. The diet should contain about 22% peNDF to maintain ruminal pH of 6,0 (Stone, 2004).

Gradual adjustment of the rumen to a high concentrate diet can help avoiding serious drop in feed intake around calving and minimize body fat mobilization in early lactation. The plasma NEFA concentration is negatively correlated with DMI and depression in feed intake around the time of calving was largely responsible for fatty liver development (Goff, 2006a). Therefore, all nutritional measures that prevent drop in DMI before parturition may be useful in prevention of fatty liver. However, it has been assumed that increased energy and nutrient density of the diet may assure maintenance of the same intake of nutrients and energy despite lower DMI around calving, and decrease rate of lipid mobilization. Increase in nutrient density during the last 2–3 weeks prepartum by increasing concentrates in the ration has been referred to as "steaming-up" or "close-up" diet (Ingvartsen, 2006). Use these diets should not last for too long because of risk of overfeeding energy and development of obesity. NRC (2001) recommends increasing the energy content of the precalving diet from 5,2 MJ/kg of DM during the "far-off" dry period, to 6,8 MJ/kg of DM for the 3 weeks before calving. This strategy is thought to prepare the cow for the metabolic demands of early lactation and thereby minimize the need for body tissue mobilization (Urton et al., 2005).

As fatty liver and ketosis in cows often occur as a complex, all measures to reduce fatty liver incidence may decrease incidence of ketosis (Grummer, 1993). Moreover, positive effects on ketosis incidence and lipid transport also have been seen when niacin or rumen-protected choline is fed to dry and fresh cows (Goff, 2006a). The treatment of pregnancy toxemia in ewes is usually unsuccessful, therefore the prevention is of key importance to reduce occurrence of the disease. Grain is a high source of available energy. Feeding 0,5-1 kg of grain daily along with high quality hay during the last four to six weeks of pregnancy will help prevent pregnancy toxemia.

Manipulation of the nutritional program of dairy cows affects rumen health, which influences hoof health (Stone, 2004). It is very evident that feeding diets that cause drop in rumen pH will result in increase in laminitis cases. To minimize the drop in rumen pH it is necessary to limit amount of concentrate fed per meal to no more than 3,6 kg and provide fresh feed at the bunk throughout the day. Diet should contain at least 25% total NDF or 19% acid detergent fiber (ADF), with non-fiber carbohydrate (NFC) levels between 35 and 40% or non-structural carbohydrate (NSC) levels between 30 to 35%, and minimize abrupt changes of ration. Adding buffers to diets may help maintain claw integrity as buffers, such as Na-bicarbonate, minimize the drop in rumen pH. Supplements that contain combinations of complexed trace minerals Zn, Mn, Cu and Co positivelly influence claw health, and it has been advised to feed them when lameness is problem in a herd.

Increased energy content of the diet fed during the prepartum period was also associated with decreased incidence of displaced abomasum (Curtis et al., 1985). Risk of displaced abomasum for cows fed silage can be almost eliminated if every cow eats a kilogram of straw daily (Ingvartsen, 2006).

## 4.2. Maintenance Of Immunity

Maintaining of a strong immune system may help to reduce incidence of infectious diseases after parturition including those related to nutritional factors: retained placenta, metritis, mastitis, laminitis, etc. Inadequate nutrition may contribute depression of the immune system that occurs around calving time (Rushen et al., 2008) when stress of parturition and metabolic challenges experienced by the dairy cow at the onset of milk production impair immune cell function (Goff, 2006a).

Dietary factor that may influence immune functions include vitamin E and Se as the most important vitamin and mineral related to immunity. To a lesser extent vitamin A and beta-carotene, Cu and Zn may play a role (Zadoks, 2006). Metabolic conditions related to parturition also may exacerbate immunosuppression, including negative energy balance (Zadoks, 2006), hypocalcaemia (Ganjkhanlou et al., 2010), and hyperketonemia (Overton & Waldron, 2004). Therefore, strategies to maintain strong immune system must include feeding adequate or increased levels of all dietary components that influence immunity, as well as prevention of metabolic disorders that may exacerbate immunosuppression.

Most measures explained in the preceding section related to adaptation of the rumen to lactation diets may help maintain immune functions throughout reduction in negative energy balance and hyperketonemia. Measures to reduce hypocalcaemia around calving are presented in the following section. Feeding vitamin E and Se have long been associated with reducing incidences of retained placentas and uterine infections. Increased protein content of the diet fed during the prepartum period was also associated with decreased incidences of retained placenta (Curtis et al., 1985).

Although feeding is generally not related to the incidence and severity of udder edema, some dietary measures can help prevention of the problem. Restriction of the salt NaCl in the diet reduces the severity of udder edema (Randalu et al., 1974). Restriction or exclusion of other Na or K sources from the diet is also recommended until the levels of dietary Na and K are not in line with dietary recommendations. NRC (2001) recommends at least 0,1% Na in dietary DM for close-up dry cow diet and at least 1% K. As membrane permeability may be a primary cause of udder edema, vitamin E supplementation precalving may be a supportive preventive measure (Thomas et al., 1990). More recently, feeding anionic diets or diets with additional antioxidants have shown some promise in reducing udder edema (Goff, 2006a). Prevention of udder edema is important as condition has temporary effects of pain and stress for the cow, and may increase risk of udder inflammation and additional health problems.

## 4.3. Prevention Of Parturient Hypocalcaemia

Concerning effects of hypocalcaemia on many physiological functions it is obvious that health, production and reproduction can be compromised in hypocalcemic cows even in the absence of clinical signs of paresis (Oetzel, 1996). Therefore, measures to prevent decline of blood Ca in cows after calving can improve milk production in herds that apparently have no problem with this disease. Prevention aims to increase the mobilization of Ca from the skeleton, or its absorption in the gastrointestinal tract, or both. The most important nutritional measures for prevention of hypocalcaemia include manipulation of minerals in the diet which is nutritionally balanced in accordance with requirements of the cows several weeks before calving. This method of prevention is based on limiting the consumption of Ca or increasing the content of anions in the diet.

### 4.3.1. Anionic Diet

An anionic, or acidic, diet is one supplemented with anionic salts to provide more anions (Cl- and S2-) relative to the cations (Na+ and K+). It has a negative dietary cation-anion difference (DCAD), calculated as mEq/kg of DM = (Na + K) – (Cl + S). It is well known that anionic diets help maintain blood Ca at parturition and prevent milk fever when fed to cows during the last several weeks of pregnancy (Block, 1984). Feeding an anionic diet before parturition has been advised if the incidence of milk fever in a herd exceeds 10%, and also when it is desired to improve the health status and production in herds in which MF is not a serious problem (Horst et al., 1994; Oetzel, 1993).

Diet with excess of anions, mainly Cl- and S2-, relative to the cations Na+ and K+ acidifies the body and is therefore considered as acidogenic (Goff & Horst, 1997b). The two PTH-dependent functions, including bone resorption and production of 1,25-(OH)2D, were enhanced after feeding acidogenic diet which contributes to lower incidence of hypocalcaemia after calving (Goff, 2000). However, the diets of dry cows in the field are always more or less alkalogenic and have a positive DCAD. When feeding diets with positive DCAD the acid-base balance tends to metabolic alkalosis, and vice versa. Formulation of acidogenic diet is only possible by supplementation with chloride and sulfate in a quantity that provides a relative excess of Cl- and S2- in the diet. Manipulation of dietary cations and anions, however, is limited by metabolic requirements and by tolerance levels of minerals in the diet.

The literature offered a dozen different formulas to calculate DCAD, which include some or all dietary minerals with or without the use of coefficients related to their utilization from the diet or the degree of influence on acid-base status:

| | |
|---|---|
| A. (Na+K) – (Cl+S) | Ender et al. (1971) |
| B. (Na+K) – (Cl+S+P) | Lomba et al. (1978) |
| C. (Na+K) – Cl | Gaynor et al. (1989) |
| D. (Na+K+Ca+0,3P) – (Cl+S) | Goff et al. (1991) |
| E. (Na+K+0,38Ca+0,3Mg) – (Cl+0,6S+0,5P) | Goff (1992) |
| F. (Na+K+0,38Ca+0,25P+0,3Mg) – (Cl-0,6S) | Goff (1992) |
| G. (Na+K+0,38Ca+0,3Mg) – (Cl-0,6S) | Horst et al. (1997) |
| H. (Na+K+0,15Ca+0,15Mg) – (Cl+0,2S+0,3P) | Oetzel (2000) |
| I. (Na+K+0,15Ca+0,15Mg) – (Cl+0,25S+0,5P) | Goff (2000) |
| J. (Na+K) – (Cl+0,6S) | Goff et al. (2004) |

Equations A, B, and C imply the complete absorption of each dietary element that can be considered accurate only in the case of equation C. Equations E and G take into account the coefficients of utilization of those elements that are not completely utilized. Equations H and I could be considered biologically and functionally most accurate because they include the degree of influence of individual ions on acid-base status, and are physiologically the most relevant (Horst & Goff, 1997; Goff, 2000).

Equation F can be used to assess the risk of developing MF. All the elements that increase the risk of MF are on the left side, while those who decrease the risk are on the right side of equation. Equation D can serve the same purpose (Goff et al., 1991). The last equation J, (Na + K) − (Cl + 0,6S), was the most recently proposed by Goff et al. (2004).

Oetzel (1991) collected data from 75 published experiments and determined the risk factors of diet in the development of MF using meta-analysis technique. Comparing the three different equations to calculate DCAD he found that the equation (Na + K) - (Cl + S) was strongly correlated with the appearance of MF. The author in the same study well justified the inclusion of S in the calculation of DCAD. Charbonneau et al. (2006) used the same technique but found that the equation (Na + K) − (Cl + 0,6S) was the most highly associated with clinical milk fever (R2 = 0,44) and urinary pH (R2 = 0,85).

**Table 1.** Mineral elements in the diet used for calculation of DCAD

| Element | $A_r$ | Valence | Equivalent weight (g/Eq) | Conversion factor (from % in mEq/kg) |
|---|---|---|---|---|
| Na | 22,99 | 1 | 22,99 | 434,98 |
| K | 39,09 | 1 | 39,09 | 255,77 |
| Ca | 40,08 | 2 | 20,04 | 499,00 |
| Mg | 24,31 | 2 | 12,16 | 822,88 |
| Cl | 35,45 | 1 | 35,45 | 282,06 |
| S | 32,06 | 2 | 16,03 | 623,83 |
| P | 30,97 | 3 | 10,32 | 968,56 |

Mineral content in feed is usually expressed as a percentage so it is necessary to convert it in mEq/kg for calculation of DCAD. Equivalent weight (g/Eq) of each element is calculated by dividing the relative atomic weight (Ar) with its valence. For example, Ar of S is 32,06. Dividing 32,06 by 2 we get the equivalent weight of S which is 16,03 g/Eq. Factor for conversion of percentage in mEq/kg is calculated by dividing the 10.000 with equivalent weight of the element. For S it is 10.000/16,03 = 623,83 (Table 1). Multiplying this factor with percentage of the element in dietary DM its content is expressed in mEq/kg DM. When this is calculated for each element in the equation, then the sum of cations is subtracted from the sum of anions and DCAD value is obtained. The example of calculating DCAD of alfalfa hay and rapeseed meal using equation (Na + K) - (Cl + S) is given in Table 2.

**Table 2.** Example of calculating DCAD in alfalfa hay and rapeseed meal

| Alfalfa hay | K | Na | Cl | S |
|---|---|---|---|---|
| Content in DM, % | 1,71 | 0,12 | 0,38 | 0,28 |
| Factor for multiplication (F) | 255,77 | 434,98 | 282,06 | 623,83 |
| Content in mEq/kg DM (% × F) | 437,37 | 52,20 | 107,18 | 174,67 |
| DCAD mEq/kg DM [(Na+K) – (Cl+S)] | 489,57 – 281,85 = + 207,72 | | | |
| Rapeseed meal | | | | |
| Content in DM, % | 1,36 | 0,10 | 0,11 | 1,25 |
| Factor for multiplication (F) | 255,77 | 434,98 | 282,06 | 623,83 |
| Content in mEq/kg DM (% × F) | 347,85 | 43,50 | 31,03 | 779,79 |
| DCAD mEq/kg DM [(Na+K) – (Cl+S)] | 391,35 – 810,82 = - 419,47 | | | |

The example in Table 2 clearly shows the difference in the alkali-acid production potential of the two feeds depending on their mineral content. A positive DCAD of alfalfa hay is result of the relatively high K content. Extremely negative DCAD of rapeseed meal is result of high S content.

Most diets commonly used for dry cows have DCAD value of +100 to +250 mEq/kg DM (Oetzel, 2000) and even up to +500 mEq/kg DM (Pehrson et al., 1999). By choosing appropriate feeds DCAD can be reduced, but not sufficiently to influence Ca metabolism. It is therefore necessary to enrich the diet with Cl and S by adding anionic salts in a quantity that provides a negative DCAD. That is, indeed, a standard diet that contains all essential nutrients required by dry cow and was further enriched with sources of Cl and S in amount that causes a mild metabolic acidosis.

Acidogenic, or anionic mineral salts are chlorides and sulfates with a high content of Cl and S, not containing Na and K. Alkalogenic salts are, on the other hand, Na and K salts that contain organic anion which is metabolized and are also called cationic salts (Na2CO3, NaHCO3, K2CO3, KHCO3, etc). They have the effect opposite to anionic salts and are undesirable in the diet of dry cows.

To formulate an anionic diet three chlorides and three sulfates are commonly used (Table 3). Two or more of them are usually added depending on mineral content of the diet. Ca and Mg salts also serve to meet requirements in these two minerals whose content in anionic diet should be higher than in standard diet (NRC, 2001). Ammonium salts, besides Cl and S, are also a source of non-protein nitrogen.

Acid production potential or acidifying activity of certain salt depends on preferential absorption of anions in relation to cations which the salt is consisted of (Goff et al., 1991). Phosphates, for example, have a weak acidifying activity as they are absorbed only slightly better than the corresponding cations (Horst et al., 1997). NaCl is neutral salt as both elements are absorbed completely and none is metabolized, so it does not affect neither DCAD nor acid-base status.

**Table 3.**Characteristics and chemical composition of anionic salts

| Salt | $M_r$ | Val | g/Eq | DM, % | % of DM | | | | | DCAD |
|---|---|---|---|---|---|---|---|---|---|---|
| | | | | | Ca | Mg | N | Cl | S | mEq/kg DM |
| $MgSO_4x7H_2O$ [1] | 246,51 | 2 | 123,26 | 48,83 | | 20,20 | | | 26,63 | -16.613 |
| $MgCl_2x6H_2O$ [2] | 203,33 | 2 | 101,67 | 46,83 | | 25,53 | | 74,46 | | -21.002 |
| $CaSO_4x2H_2O$ [3] | 172,18 | 2 | 86,09 | 79,07 | 29,44 | | | | 23,55 | -14.691 |
| $CaCl_2x2H_2O$ [4] | 147,02 | 2 | 73,51 | 75,49 | 36,11 | | | 63,88 | | -18.018 |
| $(NH_4)_2SO_4$ [5] | 132,14 | 2 | 66,07 | 100,0 | | | 21,20 | | 24,26 | -15.134 |
| $NH_4Cl$ [6] | 53,49 | 1 | 53,49 | 100,0 | | | 26,19 | 66,27 | | -18.692 |

[1]magnesium sulfate heptahydrate (Epsom salt)

[2]magnesium chloride hexahydrate

[3]calcium sulfate dihydrate (gypsum)

[4]calcium chloride dihydrate

[5]ammonium sulfate, 6amonium chloride, Mr – relative molecular weight of a substance (sum of atomic weights of all atoms in the molecule), Val – valence, g/Eq – equivalent weight (Mr/valence of cation), DCAD = (Na+K) – (Cl+S).

Anionic salts are not harmless substances and require caution for the use and handling them. The main limiting factors are bitter and salty taste and potential toxicity of higher doses. $NH4Cl$ is considered the most toxic, followed by $CaCl2$. The combination of salts is the best solution because it reduces the possibility of harmful effects (Oetzel, 2000). The most acceptable method of administration is mixing salts in the feed.

An alternative to the anionic salts is adding mineral acids directly into a diet. Very potent in causing systemic acidosis is hydrochloric acid (HCl), while sulfuric acid ($H2SO4$) is less efficient (Goff &Horst, 1997a). An additional advantage of mineral acids is that cows prefer diets with acidic taste rather than bitter-salty taste of anionic salts. Goff & Horst (1998) found that efficient prevention of MF can be achieved with only 1,5 Eq of HCl in the diet, and that with no adverse consequences up to 2,5 Eq of the acid can be fed daily. Moreover, the addition of HCl in the diet had, unlike anionic salts, a beneficial effect on the consumption of feed. For safety, it is not advisable to use pure acid or keep it on farm, but rather as commercial products in the form of acidified feeds where the acid is mixed with some feed as a carrier. Byproducts of fermentation are usually used for this purpose, and also soybean meal or sugar beet pulp (Goff & Horst, 1997a, 1998).

Feeding anionic diet few weeks before calving significantly reduces the incidences of MF and subclinical hypocalcaemia in the herd, and improves milk production and reproductive performances of cows in the subsequent lactation (Beede et al., 1991; Block, 1984; DeGroot et al., 2010). Metabolic acidosis caused by anionic diet is subclinical, mild and compensated, and with no significant impact on animal health, but may influence some physiological functions if lasts long enough. Increased sensitivity of PTH receptors in renal tissue contributes to increased production of 1,25-(OH)2D before parturition (Gaynor et al., 1989; Goff et al., 1991). Cows fed anionic diet have a higher content of 1,25-(OH)2D in plasma, although changes in secretion of PTH does not occur (Phillippo et al., 1994). The active form of vitamin D takes part in osteolysis with PTH (Horst et al., 1994; Horst, 1986) and stimulates Ca reabsorption in renal tubules (Goff, 1992). Since 1,25-(OH)2D is important for the functioning of both Ca compensation mechanisms, change in its secretion is considered as an important effect of anionic diet.

Oetzel (1993) recommended the application of anionic salts in the herds with high incidence of MF (>10%) if not possible to formulate low Ca diet, or to improve production and reproduction in the herd which apparently does not have high incidence of MF. Adding anions in the diet is, therefore, to be considered not only for preventing MF, but also for prevention of subclinical hypocalcaemia which is responsible for the frequent occurrence of other metabolic and reproductive disorders in puerperium (Horst et al., 1994). Beede (1992) states that even in well-kept herds, where cows are in proper condition with no major problems with metabolic disorders, an additional 250-500 kg of milk can be achieved by using anionic salts in diets before parturition probably because lower incidence of subclinical hypocalcaemia.

Based on results of earlier studies, Horst et al. (1997) found that the most effective diets for prevention of MF are those with DCAD from -50 to -100 mEq/kg DM, while others state that it is between -50 and -150 (Beede, 1995), or -200 (Horst et al., 1994). However, Beede et al. (1991) achieved a good preventive effect with DCAD -250 mEq/kg, and Goff et al. (1991) with -230 mEq/kg. Too low DCAD, irrespective of the amount of added salts, can reduce feed consumption because too strong metabolic acidosis which depresses the appetite in animals (Goff & Horst, 1997b). Preventive effect of anionic diet on MF incidence occurs only if DCAD is below -40 mEq/kg DM in most cases, regardless of the type and quantity of added salts.

A risk for many health and reproductive disorders in puerperium is increased in cows with hypocalcaemia (Curtis et al., 1985; Massey et al., 1993). The prevention of hypocalcaemia can, in addition to better health status of cows in puerperium, improve milk production and increase reproduction performances in the herd (Beede et al., 1991; Block, 1984; DeGroot et al., 2010). Comparing the milk production in healthy cows with those who suffered MF, Block (1984) found greater milk production by about 14% in healthy animals. The same author found that cows fed anionic diet in the dry period produced on average 486 kg (7,3%) more milk than those fed cationic diet. Beede et al. (1991) conducted a field experiment with 510 cows and showed three positive effects of anionic diet: enhanced Ca metabolism in periparturient period, increased milk production in subsequent lactation, and better reproductive performances. Cows fed anionic diet before parturition produced 327 kg more milk (about 3,6%) than control cows.

Because of unpleasant taste, anionic salts can have depressing effect on feed intake (Joyce et al., 1997; Vagnoni & Oetzel, 1998), but this is minimal if salts are properly used (Block, 1984; Moore et al., 2000; Oetzel et al., 1988). Up to 300 mEq of anions/kg DM (3-3,5 Eq of salts per day per cow) can be added to the diet without depressing effect on feed intake (Horst et al., 1994; Oetzel, 2000). Intake also could be influenced by the type of salt because more acidic salts have less favorable taste. Feed intake depression was at least expressed with $MgSO_4$, and much more with other salts, mainly $CaCl_2$ and $NH_4Cl$ (Oetzel & Barmore, 1993). Last two also have much higher acidifying activity in relation to $MgSO_4$.

The best and easiest method of application of anionic salts is mixing in TMR which effectively masks their taste. In the conventional feeding system application of anionic salts is difficult, but still possible. The salts should be manually mixed in silage or haylage, or in concentrates (Oetzel & Barmore, 1993). In the latter case, salts should be mixed in sufficient amount of concentrates containing palatable feeds, at least 2,3-2,6 kg (Oetzel & Barmore, 1993; Pehrson et al., 1999).

Oetzel (2000) finds that preventive effect of anionic salts can be achieved if cows consume anionic diet at least 5 days before calving, while others state that at least 10 days is necessary. Since the expected and actual date of calving are matter of discrepancies in practice, it is advisable to start feeding salts 3-4 weeks before expected parturition, at least 2-3 weeks, to ensure that most of the cows consume the diet at least 10 days (Beede 1992, 1995).

NRC (2001) gives fairly broad recommendations for the content of Ca in anionic diet for dry cows, from 0,6 to 1,5% DM, which is not difficult to formulate when anionic Ca salts are used. Anionic salts should not be fed if Ca intake is below 50 g/day (Oetzel, 1993). As hypomagnesaemia negatively affects Ca metabolism at the time of calving, many authors emphasize the importance of adequate content of Mg in dry cow diet (Sansom et al., 1983; Thilsing-Hansen et al., 2002). Recommendation has increased from earlier 0,20-0,25% to the current 0,35-0,40% DM either in standard or in anionic diet (NRC, 2001). Increasing dietary Mg up to 0,40% DM has no negative consequences so, even if not prove useful, poses no risk or any practical problem in the formulation of a diet when anionic Mg salts are used (Wang & Beede, 1992). On the other hand, the Mg level of 0,4% DM has been also set as the maximum tolerable (NRC, 2001).

If we assume that DCAD in efficient anionic diet should be around -100 mEq/kg DM, it means that the initial DCAD should not be greater than +200 to +250 (Horst et al., 1997). Therefore, the diet must be formulated using feeds with lower content of K, such as corn silage, grass hay, etc. If necessary, part of the forage with high content of K can be replaced with concentrates rich in fiber, such as brewers' grains, sugar beet pulp, malt sprouts, etc. The most common problem is grass silage and alfalfa silage because they can contain more than 3% K. Most cereal grains have almost neutral DCAD, while it is more negative only in rapeseed meal and brewers grains primarily due to high content of S (NRC, 2001).

In field conditions, a marked decrease in feed intake may occur when anionic salts are included in the diet for the first time, therefore, adjusting cows should last at least three days (Oetzel, 2000)

The most accurate biological indicator of the degree of acidification of the body is pH of urine (Vagnoni & Oetzel, 1998). Urinary pH in cows was decreased linearly with decreasing DCAD (Charbonneau et al., 2006; Ganjkhanlou et al., 2010). The optimum pH of urine to prevent puerperal hypocalcaemia in Holsteins fed anionic diet is 6 - 6,5 (Horst & Goff, 1997) or wider, 6 - 7 (Moore et al., 2000). In Jersey cows, however, the pH of urine is necessary to reduce to 5,8 - 6,2. If the pH is lower than the specified, amount of anionic salts should be reduced and vice versa.

### 4.3.2. Low Calcium Diet

Maintenance requirements of the cow and requirements of fetus during high pregnancy are usually satisfied with 35-45 g Ca/day (NRC, 2001). The content of Ca in the diet is often beyond that, and the requirements in this case are met mainly by passive absorption, while mechanisms of mobilization of body reserves and active absorption of Ca in the intestine are suppressed. Inactivity of these mechanisms in the dry period makes cows difficult to adapt to the sudden loss of Ca in colostrum at the moment of mammary gland activation. A few days required for starting these mechanisms causes temporary crisis in Ca homeostasis that results in decline of Ca in the blood (Goff, 1992).

When Ca content in the diet is below the minimum requirement, the animal is dependent on Ca mobilization from bone with simultaneous increase of active Ca absorption in the gut (Horst, 1986). These mechanisms can be activated before parturition and maintained active by constant stimulation up to the critical moment of parturition. Restriction of Ca intake to below 20 g/day caused negative balance of Ca and stimulates the secretion of PTH which increases tubular Ca reabsorption, bone resorption and production of 1,25-(OH)2D (Goff, 1992; Horst et al., 1997). This allows the cow to use Ca more efficiently from the diet and from body reserves immediately after calving (Goff, 1992). Introduction of low Ca diet leads to a slight decrease of Ca and P concentrations in plasma which returns to the base level in 3-4 days due to simultaneous increase in secretion of PTH that remained increased up to calving time (Goings et al., 1974). Prolonged exposure of the tissues to elevated PTH levels after feeding low Ca diet can overcome tissue resistance to PTH that might be induced by high dietary K (Goff, 2006a). This is the way to reduce or avoid period of adaptation to sudden losses of Ca which usually takes several days (Goff, 1992). Rations formulated to contain less than 20 grams of Ca proved to be very effective in preventing MF and significantly reduce the occurrence of hypocalcaemia too (Goings et al., 1971, 1974; Wiggers et al., 1975). For the best effect, restrictions of Ca consumption in the dry period should last at last 7-10 days (Goings et al., 1974; Wiggers et al., 1975) and be folowed by a high Ca diet immediately after calving (Horst et al., 1994; Oetzel, 2000). Cows fed in this way at the moment of calving mobilize about 10 g of Ca from the skeleton daily that is enough to prevent the occurrence of MF (Wiggers et al., 1975). Increasing Ca in the diet above daily needs and manipulation of its content within these limits does not affect the blood Ca status if other dietary minerals are maintained unchanged (Goff & Horst, 1997b).

Feeding low Ca diet to dry cows is considered the traditional way of preventing MF. For this reason, the use of feeds rich in Ca is usually avoided and the diet is formulated with corn silage and grass hay. Consumption of Ca can be reduced to 50-60 g/day in this way what often gives good results in the field (Goff, 1992; Horst et al., 1997). However, significant stimulation of parathyroid gland and complete preventive effect can be achieved only if Ca consumption is restricted to below 20 g/day. The formulation of such diet is, unfortunately, unrealistic in practice and that is the reason for its limited application (Goff, 1992). This type of diet has no adverse effects on production in the subsequent lactation (Goings et al., 1974), but the problem is, in addition to impracticality, that it may not last longer than 2-3 weeks because negative balance of Ca can be too exhausting for body reserves (Van Saun & Sniffen, 1996).

In the strategy for prevention of MF, Horst et al. (1997) recommended measures depending on the baseline cation-anion difference of the diet (Fig. 1). If it is less than 250 mEq/kg DM, the use of chloride and sulfate is justified without danger of low feed intake. If this value is above 250 mEq/kg DM it is necessary to consider other preventive measures, such as low Ca diet, short-term administration of oral Ca salts or some pharmacologic therapy (vitamin D analogs and active metabolites, PTH injections, etc.).

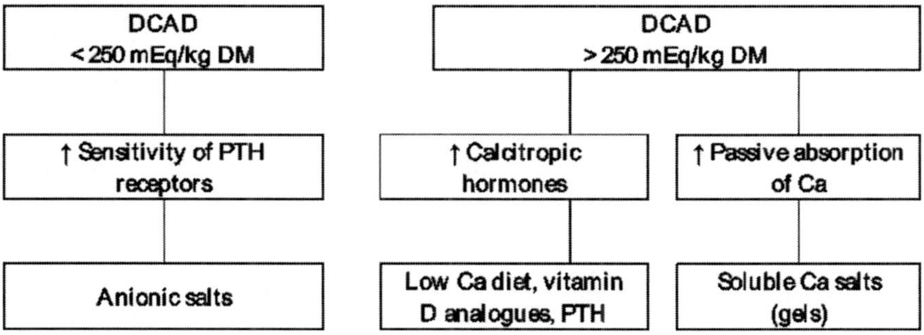


**Figure 1.** Strategy for preventing milk fever (adapted from Horst et al., 1997)

Thilsing-Hansen et al. (2002) rewieved research conducted over the past 50 years and concluded that any measures for preventing MF, even if used under ideal conditions, rarely reach preventive effect of 100%. The same authors calculated that efficiency of low Ca diets is best, reaching 80-100%, and anionic diets efficiency was 65-80%.

## 4.4. Prevention Of Hypomagnesaemia

Commonly used parameter to characterize the grass tetany potential of forage is the ratio [K/(Ca+Mg)] (Mayland, 1988). Forages containing less than 0,2% Mg and a "tetany ratio" [K/(Ca+Mg)] greater than 2,2 have increased risk of inducing grass tetany (Crawford et al., 1998). The fertilizers containing N and K are the most important factors increasing [K/(Ca+Mg)] ratio in forages. Key step in nutritional prevention of hypomagnesaemia is finding a forage with lower ratio [K/(Ca+Mg)], or modify this ratio in the diet by adding more Ca and Mg supplements. Commercial grade MgO, $MgCl_2$, $MgCO_3$ and $MgSO_4$ are good sources. As hypomagnesaemia may induce hypocalcaemia, including Ca in supplements to prevent Mg tetany may have a beneficial effect (Robinson et al., 1989).

NRC (2001) recommends Mg level of 0,35-0,40% DM in the diet for dry cows, although studies showed no benefit from increasing dietary Mg above 0.2% DM when diets contain less than 3% K (Wang & Beede, 1992). Van Saun & Sniffen (1996) recommended increasing dietary Mg above 0,20% when the content of K exceeds 1,2 to 1,5% DM. However, Goff (2006a) recommends that Mg content of the close-up dry cow ration and the early lactation ration should be between 0,35 and 0,40% as insurance against the possibility that the active transport processes for Mg absorption are impaired. In pregnant cows high Ca intake decreased utilization of dietary Mg and increases its excretion in urine (Sansom et al., 1983) that can be even worsened in metabolic acidosis (Wang & Beede, 1992). So it seems logical to increase Mg content in anionic close-up diet because it usually contains more Ca.

## 4.5. Prevention Of Milk Fat Depression

As milk fat depression has been observed over a wide range of feeding situations, this problem on dairy farms remains one of the more challenging tasks within overall nutritional management of dairy cows. Diets high in concentrates and low in fiber, and diets supplemented with plant or fish oil are the most often associated with MFD, but many other dietary factors also can affect milk fat synthesis including those which are able to alter rumen environment and those related to supply of polyunsaturated fatty acids (PUFA). Factors that can alter rumen environment include low level of physically effective NDF, feed particle size, total fiber in the diet, starch and non-structural carbohydrates, feeding pattern etc. On the other side, factors related to supply of PUFA are variation in fat content and fatty acid composition of feed ingredients, amount and availability of PUFA, and also feeding pattern.

Factor that alter rumen environment are first to consider in nutritional strategy to prevent or solve the problem of MFD, and those who affect rumen pH are the most important. Lower then normal ruminal pH, even without signs of acidosis, causes the change in bacterial population favoring those that have alternative pathways of biohydrogenation of dietary fatty acids. Inclusion of adequate level of so-called ''effective fiber'' and appropriate buffers in lactation diets can prevent drop in ruminal pH and markedly decrease MFD (Bergen, 2009). Forages, including long stemmed hay, are the main sources of effective fiber, thus the best method to maintain an adequate fat percentage in milk is to feed a balanced ration with adequate forage. The low levels of effective fiber may result from overfeeding of concentrates or the lack of forage, from consumption of large amounts of lush pasture and from silage or haylage that is too finely chopped (Perfield & Bauman, 2005).

Additional management practices to maintain a stable milk fat percentage in dairy herds include regular feeding of diet without abrupt changes, as well as feeding buffers such as Na-bicarbonate and/or Mg-oxide. Buffers are particularly useful when more than 5,5 kg of concentrate is fed per feeding and when frequent changes in diet are made.

Adding different sources of fat in dairy cow rations is a practice which has been favored when it is necessary to increase the energy consumption. However, higher amounts of fat can inhibit the activity of rumen bacteria and reduce the efficiency of fiber digestion, thus leading to a reduction in milk fat content (Schroeder et al., 2004). Although unsaturated vegetable oils can have a number of positive effects on fatty acid composition of milk fat, a negative effect on fat content must not be neglected due to the fact that the fat in milk is one of the main determinants of the price of milk and profits that farmers make. High levels of unsaturated oil in the diet can reduce fat content in milk along with possibly other adverse effects on production performances, such as drop in milk yield, decreased protein content in milk, low feed intake, etc. (Shingfield et al., 2006). Loor et al. (2002) found that the upper limit for dietary supplementation of unsaturated vegetable oils is 3,5% DM without serious nefative consequences on production. If fish oil is the supplement, it is at most 1% DM (Donovan et al., 2000).

## REFERENCES

1. A. H. Andrews, V. E. Holland-Howes, J. I. D. Wilkinson, 1996Naturally Occurring Pregnancy Toxaemia in the Ewe and Treatment with Recombinant Bovine SomatotropinSmall Ruminant Research232-3January 1997), 1911970921-4488

2. L. G. Baird, C. K. W. Muelling, 2009Risk Factors, Pathogenesis and Prevention of Subclinical Laminitis in Dairy Cows. Proceedings of the CanWest Veterinary Conference, 110Alberta, Canada, October 17-20, 2009

3. D. E. Bauman, J. M. Griinari, 2001Regulation and Nutritional Manipulation of Milk Fat: Low-Fat Milk SyndromeLivestock Production Science701-2July 2001), 15291871-1413

4. D. E. Bauman, I. I. J. W. Perfield, K. J. Harvatine, L. H. Baumgard, 2008Regulation of Fat Synthesis by Conjugated Linoleic Acid: Lactation and the Ruminant ModelJournal of Nutrition1382February 2008), 4034090022-3166

5. L. H. Baumgard, J. K. Sangster, D. E. Bauman, 2001Milk Fat Synthesis in Dairy Cows is Progressively Reduced by Increasing Supplemental Amounts of trans-10,cis-12 Conjugated Linoleic Acid (CLA).Journal of Nutrition, 1316June 2001), 176417690022-3166

6. D. K. Beede, C. Wang, G. A. Donovan, L. F. Archibald, W. K. Sanchez, 1991Dietary Cation-Anion Difference (Electrolite Balance) in Late Pregnancy. Proceedings of the 28th Florida Dairy Production Conference, 98104Gainesville, FL, April 9-10, 1991

7. D. K. Beede, 1995Macromineral Element Nutrition for the Transition Cow: Practical Implications and Strategies. Proceedings of The Tri-State Dairy Nutrition Conference, 175188Fort Wayne, IN, USA, May 11-13, 1995

8. D. K. Beede, 1992The DCAD Concept: Transition Rations for Dry Pregnant Cows. Feedstuffs, 6453December 1992), 12190014-9624

9. W. G. Bergen, 2009Milk-Fat Depression and Lipid Repartitioning in Lactating Dairy CowsJournal of Nutrition1395May 2009), 8268270022-3166

10. S. J. Bertics, R. R. Grummer, C. Cadorniga-Valino, E. E. Stoddard, 1992Effect of Prepartum Dry Matter Intake on Liver Triglyceride Concentration and Early Lactation.Journal of Dairy Science757July 1992), 191419220022-0302

11. E. Block, 1984Manipulation Dietary Anions and Cations for Prepartum Dairy Cows to Reduce Incidence of Milk Fever. Journal of Dairy Science, 6712December 1984), 293929480022-0302

12. G. Bobe, J. W. Young, D. C. Beitz, 2004Invited Review: Pathology, Etiology, Prevention, and Treatment of Fatty Liver in Dairy Cows.Journal of Dairy Science8710October 2004), 310531240022-0302

13. E. Bramley, I. J. Lean, W. J. Fulkerson, M. A. Stevenson, A. R. Rabie, N. D. Costa, 2008The Definition of Acidosis in Dairy Herds Predominantly Fed on Pasture and ConcentratesJournal of Dairy Science911January 2008), 3083210022-0302

14. E. Charbonneau, D. Pellerin, G. R. Oetzel, 2006Impact of Lowering Dietary Cation-Anion Difference in Nonlactating Dairy Cows: A

Meta-Analysis.Journal of Dairy Science892February 2006), 5375480022-0302

15. R. J. Crawford Jr, M. D. Massie, D. A. Sleper, H. E. Mayland, 1998Use of an Experimental High-Magnesium Tall Fescue to Reduce Grass Tetany in Cattle. Journal of Production Agriculture114October-December 1998), 4914960890-8524

16. C. Crnkic, S. Muratovic, S. Piplica, A. Kavazovic, S. Kutlaca, 2010Blood Plasma Mineral Profile and Health Status in Postpartum Cows Fed an Anionic Diet Before ParturitionTurkish Journal of Veterinary and Animal Science, 343June 2010), 2552601300-0128

17. C. R. Curtis, H. N. Erb, C. J. Sniffen, R. D. Smith, D. S. Kronfeld, 1985Path Analysis of Dry Period Nutrition, Postpartum Metabolic and Reproductive Disorders, and Mastitis in Holstein Cows.Journal of Dairy Science689September 1985), 234723600022-0302

18. C. R. Curtis, H. N. Erb, C. J. Sniffen, R. D. Smith, 1984Epidemiology of Parturient Paresis: Predisposing Factors with Emphasis on Dry Cow Feeding and Management.Journal of Dairy Science674April 1984), 8178250022-0302

19. M. A. De Groot, E. Block, P. D. French, 2010Effect of Prepartum Anionic Supplementation on Periparturient Feed Intake, Health, and Milk ProductionJournal of Dairy Science9311November 2010), 526952790022-0302abstract)

20. M. R. Dentine, B. T. Mc Daniel, 1984Associations of Subjective Udder Edema Scores and Descriptive Trait Codes for Udder Types.Journal of Dairy Science671January 1984), 2082150022-0302

21. I. W. Dishington, 1974The Role of Age on the Induction of Hypocalcaemic Paresis Puerperalis in Dairy Cows. Nordisk Veterinaermedicin, 263March 1974), 2052100029-1579

22. D. C. Donovan, D. J. Schingoethe, R. J. Baer, J. Ryali, A. R. Hippen, S. T. Franklin, 2000Influence of Dietary Fish Oil on Conjugated Linoleic Acid and Other Fatty Acids in Milk Fat from Lactating Dairy Cows.Journal of Dairy Science8311November 2000), 262026280022-0302

23. F. Ender, I. W. Dishington, A. Helgebostad, 1971Calcium Balance Studies in Dairy Cows Under Experimental Induction and Prevention

of Hypocalcaemic Paresis PuerperalisZeitschrift fur Tierphysiologie, Tierernahrung und Futtermittelkunde, 281-5January-December 1971), 2332560044-3565

24. C. Fourichon, H. Seegers, N. Bareille, F. Beaudeau, 1999Effects of Disease on Milk Production in the Dairy Cow: A Review.Preventive Veterinary Medicine411June 1999), 1350167-5877

25. N. C. Friggens, C. Ridder, P. Løvendahl, 2007On the Use of Milk Composition Measures to Predict the Energy Balance of Dairy Cows.Journal of Dairy Science9012December 2007), 545354 670022-0302

26. M. Ganjkhanlou, A. Nikkhah, A. Zali, 2010Effect of Dietary Cation-Anion Balance on Milk Production and Blood Mineral of Holstein Cows During the Last Two Months of PregnancyAfrican Journal of Biotechnology936September 2010), 598359881684-5315

27. P. J. Gaynor, F. J. Mueller, J. K. Miller, N. Ramsey, J. P. Goff, R. L. Horst, 1989Parturient Hypocalcaemia in Jersey Cows Fed Alfalfa Haylage Based Diets With Different Cation to Anion Ratios. Journal of Dairy Science, 7210October 1989), 252525310022-0302

28. J. P. Goff, R. L. Horst, F. J. Mueller, J. K. Miller, G. A. Kiess, H. H. Dowlen, 1991Addition of Chloride to a Prepartal Diet High in Cations Increases 1,25-dihydroxyvitamin D Response to Hypocalcaemia Preventing Milk FeverJournal of Dairy Science 7411November 1991), 386338710022-0302

29. J. P. Goff, R. Ruiz, R. L. Horst, 2004Relative Acidifying Activity of Anionic Salts Commonly Used to Prevent Milk Fever.Journal of Dairy Science875May 2004), 124512550022-0302

30. J. P. Goff, 1992Cation-Anion Difference of Diets and Its Influence on Milk Fever and Subsequent Lactation: the Good and the Bad News. Proceedings of Cornell Nutrition Conference For Feed Manufacturers, 1481590885-7687Rochester, N.Y, USA, October 13-15, 1992

31. J. P. Goff, 2000Pathophysiology of Calcium and Phosphorus Disorders.Veterinary Clinics of North America: Food Animal Practice, 162July 2000), 3193370749-0720

32. J. P. Goff, 2006aMajor Advances in Our Understanding of Nutritional Influences on Bovine Health.Journal of Dairy Science894April 2006), 129213010022-0302

33. J. P. Goff, 2006bMineral Disorders of the Transition Period: Origin And Control. Proceedings of XXIV World Buiatrics Congress, 192-90362-340-7France, October 15-19, 2006

34. J. P. Goff, R. L. Horst, 1997aComparison of Sulfuric Acid with Hydrochloric Acid as a Source of Acidifying Anions. Journal of Animal Science, 75Suppl.1, (January 1997), 980021-8812

35. J. P. Goff, R. L. Horst, 1997bEffects of the Addition of Potassium or Sodium, but not Calcium, to Prepartum Rations on Milk Fever in Dairy CowsJournal of Dairy Science801January 1997), 1761860022-0302

36. J. P. Goff, R. L. Horst, 1997cPhysiological Changes at Parturition and their Relationship to Metabolic Disorders.Journal of Dairy Science807July 1997), 126012680022-0302

37. J. P. Goff, R. L. Horst, 1998Use of Hydrochloric Acid as a Source of Anions for Prevention of Milk Fever.Journal of Dairy Science8111November 1998), 287428800022-0302

38. R. L. Goings, N. L. Jacobson, E. T. Littledike, 1971Responses of Cows to a Very Low Calcium Intake at Parturition. Journal of Dairy Science, 545May 1971), 7917920022-0302

39. R. L. Goings, N. L. Jacobson, D. C. Beitz, E. T. Littledike, K. D. Wiggers, 1974Prevention of Parturient Paresis by a Prepartum Calcium-Deficient Diet.Journal of Dairy Science5710October 1974), 118411880022-0302

40. L. E. Green, V. J. Hedges, Y. H. Schukken, R. W. Blowey, A. J. Packington, 2002The Impact of Clinical Lameness on the Milk Yield of Dairy Cows.Journal of Dairy Science859September 2002), 225022560022-0302

41. R. R. Grummer, 1993Etiology of Lipid-Related Metabolic Disorders in Periparturient Dairy Cows.Journal of Dairy Science7612 December 1993), 388238960022-0302

42. C. L. Guard, 1996Fresh Cow Problems are Costy; Culling Hurts the Most. Hoard's Dairyman, 14180018-2885

43. K. J. Harvatine, Y. R. Boisclair, D. E. Bauman, 2009Recent Advances in the Regulation of Milk Fat SynthesisAnimal31January 2009), 40541751-7311

44. R. L. Horst, 1986Regulation of Calcium and Phosphorus Homeostasis in Dairy Cow. Journal Of Dairy Science692February 1986), 6046160022-0302

45. R. L. Horst, J. P. Goff, 1997Milk Fever and Dietary Potassium. Proceedings of Cornell Nutrition Conference for Feed Manufacturers, 1811890885-7687Rochester, NY, USA, October 21-23, 1997

46. R. L. Horst, J. P. Goff, T. A. Reinhardt, D. R. Buxton, 1997Strategies for Preventing Milk Fever in Dairy Cattle.Journal of Dairy Science807July 1997), 126912800022-0302

47. K. L. Ingvartsen, 2006Feeding- and Management-Related Diseases in the Transition Cow: Physiological Adaptations Around Calving and Strategies to Reduce Feeding-Related Diseases. Animal Feed Science and Technology, 1263-4March 2006), 1752130377-8401

48. E. H. Johnson, K. Al-Habsi, E. Kaplan, A. Srikandakumar, I. T. Kadim, K. Annamalai, R. Al-Busaidy, O. Mahgoub, 2004Caprine Hepatic Lipidosis Induced Through the Intake of Low Levels of Dietary CobaltThe Veterinary Journal1682September 2004), 1741791090-0233

49. P. W. Joyce, W. K. Sanchez, J. P. Goff, 1997Effect of Anionic Salts in Prepartum Diets Based on Alfalfa.Journal of Dairy Science8011November 1997), 286628750022-0302

50. D. S. Kronfeld, S. Donoghue, R. L. Copp, F. M. Stearns, R. H. Engle, 1982Nutritional Status of Dairy Cows Indicated by Analysis of Blood.Journal of Dairy Science6510October 1982), 192519330022-0302

51. G. S. Lewis, 1997Uterine Health and Disorders.Journal of Dairy Science805May 1997), 9849940022-0302

52. D. N. Logue, J. E. Offer, 2001Guest Editorial: The Effect of Forage Type on Foot Health in Dairy Heifers. The Veterinary Journal, 1621July 2001), 781090-0233

53. F. Lomba, E. T. Chauvaux, L. Lengele, V. Bienfet, 1978Calcium Digestibility in Cows as Influenced by the Excess of Alkaline Ions Over Stable Acid Ions in Their Diets.The British Journal of Nutrition393May 1978), 4254290007-1145

54. J. J. Loor, L. E. Quinlan, A. B. P. A. Bandara, J. H. Herbein, 2002Distribution of trans-vaccenic acid and cis-9,trans-11-conjugated linoleic acid (rumenic acid) in blood plasma lipid fractions and secretion in milk fat of Jersey cows fed canola or soybean oil. Animal Research, 512March-April 2002), 1191341627-3583

55. P. V. Malven, R. E. Erb, M. F. D'amico, T. S. Stewart, B. P. Chew, 1983Factors Associated with Edema of the Mammary Gland in Primigravid Dairy Heifers.Journal of Dairy Science662February 1983), 2462520022-0302

56. C. D. Massey, C. Wang, G. A. Donovan, D. K. Beede, 1993Hypocalcaemia at Parturition as a Risk Factor for Left Displacement of the Abomasum in Dairy Cows. Journal of the American Veterinary Medical Association, 2036September 1993), 8528530003-1488

57. H. F. Mayland, 1988Grass Tetany. In: The Ruminant Animal; Digestive Physiology and Nutrition, D.C. Church, (Ed.), 511531Prentice Hall, 0-88133-740-4Cliffs, N.J., USA

58. J. J. Michal, L. R. Heirman, T. S. Wong, B. P. Chew, M. Frlgg, L. Volker, 1994Modulatory Effects of Dietary Beta Carotene on Blood and Mammary Leukocyte Function in Periparturient Dairy Cows.Journal of Dairy Science775May 1994), 140814210022-0302

59. S. J. Moore, M. J. Van de Haar, B. K. Sharma, T. E. Pilbeam, D. K. Beede, H. F. Bucholtz, J. S. Liesman, R. L. Horst, J. P. Goff, 2000Effects of Altering Dietary Cation-Anion Difference on Calcium and Energy Metabolism in Peripartum Cows.Journal of Dairy Science839September 2000), 209521040022-0302

60. M. Morgante, 2004Digestive Disturbances and Metabolic-Nutritional Disorders.In: Dairy sheep nutrition. G. Pulina (Ed.), 165191CAB International, 0-85199-681-7UK.

61. K. E. Nestor Jr, R. W. Hemken, R. J. Harmon, Influence of Sodium Chloride and Potassium Bicarbonate on Udder Edema and Selected

Blood Parameters.Journal of Dairy Science712February 198836637 20022-0302

62. NRC,2001Nutrient Requirement Of Dairy Cattle.th rev.ed., National Academy Press, 0-30906-997-1DC, USA

63. G. R. Oetzel, 1988Parturient Paresis and Hypocalcaemia in Ruminant Livestock. Veterinary Clinics of North America: Food Animal Practice, 42July 1988), 3515640749-0720

64. G. R. Oetzel, 1991Meta-Analisis of Nutritional Risk Factor for Milk Fever in Dairy Cattle. Journal of Dairy Science, 7411November 1991), 390039120022-0302

65. G. R. Oetzel, 1993Use of Anionic Salts for Prevention Milk Fever in Dairy Cattle. Compendium On Continuing Education For The Practicing Veterinarian, 158August 1993), 113811460193-1903

66. G. R. Oetzel, 1996Effect of Calcium Chloride Gel Treatment in Dairy Cows On Incidence of Periparturient Diseases.Journal of the American Veterinary Medical Association2095September 1996), 9589610003-1488

67. G. R. Oetzel, 2000Management of Dry Cows for the Prevention of Milk Fever and Other Mineral Disorders.Veterinary Clinics of North America: Food Animal Practice, 162July 2000), 3693860749-0720

68. G. R. Oetzel, J. A. Barmore, 1993Intake of a Concentrate Mixture Containing Various Anionic Salts Fed to Pregnant, Nonlactating Dairy Cows.Journal of Dairy Science766Juny 1993), 161716230022-0302

69. G. R. Oetzel, J. D. Olson, C. R. Curtis, M. J. Fettman, 1988Ammonium Chloride and Ammonium Sulfate for Prevention of Parturient Paresis in Dairy Cows.Journal of Dairy Science 7112December 1988), 330233090022-0302

70. T. R. Overton, M. R. Waldron, 2004Nutritional Management of Transition Dairy Cows: Strategies to Optimize Metabolic HealthJournal of Dairy Science87E. Suppl.), (July 2004), E105E1190022-0302

71. B. Pehrson, 1996Milk Analysis as an Indicator of the Nutritional and Disease Status of Dairy Cow. In: Recent Advances in Animal

Nutrition, P.C. Garnsworthy, J. Wiseman, W. Haresign, (Eds.), 113129Nottingham University Press, 1-89767-603-4

72. B. Pehrson, C. Svensson, I. Gruvaeus, M. Virkki, 1999The Influence of Acidic Diets on the Acid-Base Balance of Dry Cows and the Effect of Fertilization on the Mineral Content of Grass.Journal of Dairy Science826June 1999), 13101316 0022-0302

73. I. I. J. W. Perfield, D. E. Bauman, 2005Current Theories and Recent Advances in the Biology of Milk Fat Depression. Proceedings of Cornell Nutrition Conference for Feed Manufacturers, 951069-99825-829-4

74. M. Phillippo, G. W. Reid, I. M. Nevison, 1994Parturient Hypocalcaemia in Dairy Cows: Effects of Dietary Acidity on Plasma Minerals and Calciotrophic Hormones.Research in Veterinary Science563May 1994), 3033090034-5288

75. O. M. Radostits, C. C. Gay, D. C. Blood, K. W. Hinchcliff, 2000Veterinary Medicine: A Textbook of the Diseases of Cattle, Sheep, Pigs, Goats and Horses.th Edition. 14171476WB Saunders, 0-70202-604-2UK

76. C. J. Ramberg Jr, E. K. Johnson, R. D. Fargo, D. S. Kronfeld, 1984Calcium Homeostasis in Cows with Special Reference to Parturient Hypocalcaemia. American Journal of Physiology, 2465May 1984), R698R7040363-6119

77. W. E. Randalu, R. W. Hemken, L. S. Bull, L. W. Douglas, 1974Effect of Dietary Sodium and Potassium on Udder Edema in Holstein Heifers. Journal Of Dairy Science, 574April 1974), 4724750022-0302

78. D. L. Robinson, L. C. Kappel, J. A. Boling, 1989Management Practices to Overcome the Incidence of Grass Tetany.Journal of Animal Science6712December 1989), 347034840021-8812

79. J. Rushen, A. M. de Passillé, Keyserlingk. M. A. G. von, D. M. Weary, 2008The Welfare of CattleAnimal Welfare Series, 5C.Phillips (Ed.), Springer, 978-1-40206-558-3Dordrecht, The Netherlands

80. B. F. Sansom, R. Manston, M. J. Vagg, 1983Magnesium and Milk FeverVeterinary Record11219May 1983), 4474492042-7670

81. C. Schlumbohm, J. Harmeyer, 2003Hypocalcaemia Reduces Endogenous Glucose Production in Hyperketonemic Sheep. Journal of Dairy Science, 866June 2003), 195319620022-0302

82. G. H. Schmidt, L. H. Schultz, 1959Effect of Three Levels of Grain Feeding During the Dry Period on the Incidence of Ketosis, Severity of Udder Edema, and Subsequent Milk Production of Dairy Cows.Journal Of Dairy Science421January 1959), 1701790022-0302

83. J. T. Schonewille, H. Everts, S. Jittakhot, A. C. Beynen, 2008Quantitative Prediction of Magnesium Absorption in Dairy CowsJournal of Dairy Science911January 2008), 2712780022-0302

84. G. F. Schroeder, G. A. Gagliostro, F. Bargo, J. E. Delahoy, L. D. Muller, 2004Effects of Fat Supplementation on Milk Production and Composition by Dairy Cows on Pasture: A ReviewLivestock Production Science861-3March 2004), 1181871-1413

85. R. D. Shaver, 1997Nutritional Risk Factors in the Etiology of Left Displaced Abomasum in Dairy Cows: A Review.Journal of Dairy Science8010October 1997), 244924530022-0302

86. C. Stelletta, M. Gianesella, M. Morgante, 2008Metabolic and Nutritional Diseases.In: Dairy Goats Feeding And Nutrition. A.Cannas, G.Pulina (Eds.), 263288CAB International, 1-84593-348-6UK

87. W. C. Stone, 2004Nutritional Approaches to Minimize Subacute Ruminal Acidosis and Laminitis in Dairy CattleJournal of Dairy Science87E. Suppl., (July 2004), E13E260022-0302

88. T. Thilsing-Hansen, R. J. Jorgensen, S. Ostergaard, 2002Milk Fever Control Principles: A Review.Acta Veterinaria Scandinavica431 March 2002), 1191751-0147

89. D. G. Thomas, J. K. Miller, F. J. Mueller, F. C. Madsen, 1990Udder Edema Reduced by Prepartum Vitamin E Supplementation. Journal of Dairy Science, 73Suppl.1, 2710022-0302

90. G. Urton, Keyserlingk. M. A. G. von, D. M. Weary, 2005Feeding Behavior Identifies Dairy Cows at Risk for Metritis.Journal of Dairy Science888August 2005), 284328490022-0302

91. D. B. Vagnoni, G. R. Oetzel, 1998Effect of Dietary Cation-Anion Difference on the Acid-Base Status of Dry Cows. Journal of Dairy Science816June 1998), 164316520022-0302

92. R. J. Van Saun, C. J. Sniffen, 1996Nutritional Management of the Pregnant Dairy Cow to Optimize Health, Lactation and Reproductive PerformanceAnimal Feed Science and Technology591-3June 1996), 13260377-8401

93. C. Wang, D. K. Beede, 1992Effects of Diet Magnesium on Acid-Base Status and Calcium Metabolism of Dry Cows Fed Acidogenic Salts.Journal of Dairy Science753March 1992), 8298360022-0302

94. K. D. Wiggers, D. K. Nelson, N. L. Jacobson, 1975Prevetion of Parturient Paresis by a Low-Calcium Diet Prepartum: A Field Study. Journal of Dairy Science, 583March 1975), 4304310022-0302

95. R. N. Zadoks, 2006The Role of Nutrition in Udder Health. Proceedings of High Plains Dairy Conference, 115128Albuquerque, NM, March 17, 2006

# CHAPTER 4

# Effect of feeding different levels of palm kernel cake fermented by *Paenibacillus polymyxa* ATCC 842 on nutrient digestibility, intestinal morphology, and gut microflora in broiler chickens

**M.I. Alshelmani[a, b], T.C. Loh[a, c], H.L. Foo[d, e], A.Q. Sazili[a], W.H. Lau[f]**

[a] Department of Animal Science, Faculty of Agriculture, Universiti Putra Malaysia, 43400 Serdang, Malaysia

[b] Department of Animal Production, Faculty of Agriculture, University of Sebha, Sebha, Libya

[c] Institute of Tropical Agriculture, Universiti Putra Malaysia, 43400 Serdang, Malaysia

[d]Department of Bioprocess Technology, Faculty of Biotechnology and Biomolecular Sciences, Universiti Putra Malaysia, 43400 Serdang,Malaysia

[e] Institute of Bioscience, Universiti Putra Malaysia, 43400 Serdang, Malaysia

[e] Department of Plant Protection, Faculty of Agriculture, Universiti Putra Malaysia, 43400 Serdang, Malaysia

## HIGHLIGHTS

- PKC decreased nutrient digestibility in broiler chickens.
- FPKC by *Paenibacillus polymyxa* ATCC 842 improved nutrient utilization.
- FPKC improved the gut microflora in broiler chickens.
- Both PKC and FPKC did not affect the villi height at 15%.

## ABSTRACT

A feeding trial was conducted to investigate the effect of palm kernel cake fermented (FPKC) by *Paenibacillus polymyxa* ATCC 842 on nutrient digestibility, villi height, and gut microflora. A total of 245 one-day-old broiler chicks (Cobb500) were purchased from a local hatchery and raised in an open house system in the research unit at Universiti Putra Malaysia. The chicks were individually wing-banded, weighed and randomly distributed into seven groups with five replicates (pens) and seven birds per pen. The birds were fed diets containing 0, 5, 10 and 15% palm kernel cake (PKC) and 5, 10, 15% FPKC. All the birds were fed *ad libitum* during the feeding trial, which lasted 42 days. One bird and two birds per replicate in each dietary treatment were randomly selected at 3 and 6 weeks, respectively. Ileal digesta and small intestinal tissues were collected. The findings showed that feeding of 10% or 15% PKC led to a significant decrease ($P < 0.05$) in nutrient digestibility compared with the control group, whereas feeding of 10% or 15% FPKC led to a significant improvement ($P < 0.05$) in nutrient digestibility (except in nitrogen free extract at 3 weeks of age) compared with those groups of birds fed with PKC. No significant differences ($P > 0.05$) were observed between the dietary treatments in terms of intestinal morphology. However, gut microflora were improved ($P < 0.05$) in a group of birds fed with 15% FPKC compared to the other dietary treatments. In conclusion, the current experiment showed that FPKC by *P. polymyxa* ATCC 842 could be fed to broiler chickens up to 15% in their rations without any adverse effect on nutrient digestibility.

## ABBREVIATIONS

CFU - colony forming unit

ENT - enterobacteriaceae

FPKC - fermented palm kernel cake

GIT - gastrointestinal tract

LAB - lactic acid bacteria

NSPs - non-starch polysaccharides

PKC - palm kernel cake

SSF - solid state fermentation

## KEYWORDS

Palm kernel cake; *Paenibacillus polymyxa*; Nutrient digestibility; Intestinal morphology; Gut microflora

## 1. INTRODUCTION

Malaysia has an abundant amount of palm kernel cake (PKC), which is considered to be an agro-industrial waste derived from the extraction process of oil from palm fruits. However, PKC contains high levels of insoluble fibers, a coarse texture and gritty appearance (McDonald et al., 1995, O'Mara Mulligan et al., 1999 and Sundu and Dingle, 2002), and non-starch polysaccharides (NSPs), such as mannan, xylan and cellulose (Sundu and Dingle, 2002, Alimon, 2004 and Francesch and Brufau, 2004). The presence of high levels of fiber in poultry diets may lead to a decrease in the surface, width and height of intestinal villi (Moharrery and Mohammadpour, 2005 and Kalmendal et al., 2011). Thus, the nutrient utilization may be negatively affected. The effect of dietary fiber on villi height is variable and depends on the physio-chemical characteristics of the fibers, levels, duration of the dietary fiber in the diet, animal species, age (Montagne et al., 2003) and health status of the bird (Mateos et al., 2012). It is suggested that PKC could also increase the non-pathogenic bacteria, such as lactic acid bacteria (LAB) (Zulkifli et al., 2009). At the same time, the *Enterobacteriaceae* group (ENT), which is considered to be pathogens, may decrease in birds fed diets containing PKC. The colonization of LAB in the intestinal tract may be influenced by the composition of the diet ( Wenk, 2001 and Zulkifli et al., 2009). It was reported by Zulkifli et al. (2009) that the presence of 78% NSPs in PKC might have a profound effect on gut microflora. Wenk (2001) andChen et al. (2014) reported that long-term feeding of high ratios of dietary fiber could alter the physiological and anatomical characteristics of the gastrointestinal tract (GIT) of pigs. The soluble dietary fibers may increase ileal and post-ileal microbial fermentation (Wenk, 2001). The reduction of nutrient digestibility could be attributed not only to the presence of insoluble fibers, but also to the high content of soluble fibers such as NSPs. These molecules can increase the passage rate of ingesta and may decrease nutrient absorption. Consequently, high levels of dietary fiber lead to a reduction in the digestibility of energy, starch, protein and lipids in monogastrics (Montagne et al., 2003). In contrast, the digestibility of amino acids has been improved in broiler chickens when PKC was fermented with *Aspergillus wentii* TISTR 3075 ( Muangkeow and Chinajariyawong, 2009). In addition, the nutritive value of PKC fermented by cellulolytic microorganisms was improved, and it could be beneficial in terms of feed cost per kilogram diet and poultry performance (Marini et al., 2005). Because there is increasing interest to improve the nutritive quality of PKC using cellulolytic microorganisms via solid state fermentation (SSF), it seems to be vital to investigate the effect of PKC fermented by cellulolytic microbes. Therefore, the objective of the current trial was to investigate the effect of PKC fermented by *Paenibacillus polymyxa* ATCC 842 (FPKC) on nutrient digestibility, intestinal morphology and gut microflora in broiler chickens.

# 2. MATERIALS AND METHODS

## 2.1. Birds and experimental diets

The study was conducted according to the guidelines of the Research Policy on Animal Ethics of the Universiti Putra Malaysia. A total of 245 male broiler one-day-old chicks (Cobb 500) were purchased from a local hatchery. The chicks were individually weighed, wing banded and randomly allocated into 35 pens of seven chicks per pen. The birds were raised in an open house system in the Poultry Unit, Department of Animal Science, Faculty of Agriculture, Universiti Putra Malaysia (UPM). Seven dietary treatments were formulated in the feed factory at the Poultry Unit, and the PKC levels were 0%, 5%, 10%, 15%, whereas the FPKC levels were 5%, 10% and 15%. The FPKC was produced using*Paenibacillus polymyxa* ATCC 842 under SSF for nine days ( Alshelmani et al., 2014). Each treatment consisted of five replicates, and the feeding trial lasted for 42 days. The birds had access to the feed and drinking water *ad libitum*. The birds were vaccinated against infectious bronchitis disease and Newcastle disease, and Gumboro (infectious bursal) disease at 7 and 21 days, respectively. The starter ( Table 1) and finisher (Table 2) diets were offered to the chickens from 0 to 21 days and 22 to 42 days, respectively. All the feed were formulated as isocaloric and iso-nitrogenous based on NRC (1994)recommendation, and the nutrient content of PKC and FPKC were considered for feed formulation. The birds were raised in the conventional open-sided house with cyclic temperature (maximum 35 °C and minimum 24 °C) and humidity (maximum 91% and minimum 65%). Diets were formulated using FeedLIVE software (FeedLIVE 1.52, Thailand). Dry matter (DM), CP, ash, and ether extract (EE) were determined using procedure 934.01, 976.05, 942.05, 920.39 of AOAC (1990), respectively, while acide tetergent fiber (ADF) and neutral detergent fiber (NDF) were determined based on the procedure of Goering and Van Soest (1970).

## 2.2. Samples and data collection

One bird and two birds per replicate in each dietary treatment were randomly selected at 3 and 6 weeks, respectively. The birds were slaughtered by *halal* neck cut for sampling of the ileal digesta ( Ahmed et al., 2014) and small intestines for further analysis. The ileal digesta was collected from the Meckel's diverticulum to 1 cm before ileo-ceco junction. The ileal digesta samples were taken in order to determine the lactic acid bacteria (LAB) and *Enterobacteriaceae* (ENT) counts, as well as nutrient digestibility. The small intestine samples were collected for measuring villi height and crypt depth.

**Table 1.** Ingredient and chemical compositions of broiler starter diet.

| Ingredient (%) | PKC[1] level (%) | | | | FPKC[2] level (%) | | |
|---|---|---|---|---|---|---|---|
| | 0 | 5 | 10 | 15 | 5 | 10 | 15 |
| Yellow corn | 55.94 | 50.05 | 44.51 | 39.15 | 50.05 | 44.51 | 39.15 |
| Palm oil | 3.46 | 5.16 | 6.73 | 8.24 | 5.16 | 6.73 | 8.24 |
| PKC | 0.00 | 5.00 | 10.00 | 15.00 | 0.00 | 0.00 | 0.00 |
| FPKC | 0.00 | 0.00 | 0.00 | 0.00 | 5.00 | 10.00 | 15.00 |
| Soybean meal 44% CP | 31.95 | 30.71 | 29.62 | 28.54 | 30.71 | 29.62 | 28.54 |
| Fish meal 55% CP | 5.00 | 5.00 | 5.00 | 5.00 | 5.00 | 5.00 | 5.00 |
| DCP[3] 18% | 1.60 | 1.80 | 1.80 | 1.70 | 1.80 | 1.80 | 1.70 |
| Calcium carbonate | 0.30 | 0.37 | 0.37 | 0.37 | 0.37 | 0.37 | 0.37 |
| DL-Methionine | 0.20 | 0.28 | 0.28 | 0.25 | 0.28 | 0.28 | 0.25 |
| L-Lysine | 0.00 | 0.10 | 0.16 | 0.22 | 0.10 | 0.16 | 0.22 |
| Salt | 0.32 | 0.30 | 0.30 | 0.30 | 0.30 | 0.30 | 0.30 |
| Vitamin Premix[a] | 0.50 | 0.50 | 0.50 | 0.50 | 0.50 | 0.50 | 0.50 |
| Mineral Premix[b] | 0.50 | 0.50 | 0.50 | 0.50 | 0.50 | 0.50 | 0.50 |
| Toxin binder | 0.14 | 0.14 | 0.14 | 0.14 | 0.14 | 0.14 | 0.14 |
| Antioxidant[c] | 0.01 | 0.01 | 0.01 | 0.01 | 0.01 | 0.01 | 0.01 |
| Choline chloride | 0.08 | 0.08 | 0.08 | 0.08 | 0.08 | 0.08 | 0.08 |
| Total | 100.00 | 100.00 | 100.00 | 100.00 | 100.00 | 100.00 | 100.00 |
| Calculated analyses[d] | | | | | | | |
| ME[6] (kcal/kg) | 3000.10 | 3000.03 | 3000.14 | 3000.71 | 3000.03 | 3000.14 | 3000.71 |
| Methionine + Cystine (%) | 0.94 | 1.00 | 0.99 | 0.95 | 1.00 | 1.00 | 0.96 |
| Lysine (%) | 1.19 | 1.25 | 1.27 | 1.30 | 1.25 | 1.27 | 1.30 |
| Arginine (%) | 1.43 | 1.47 | 1.50 | 1.54 | 1.47 | 1.50 | 1.54 |
| Lys: Arg ratio | 1:1.2 | 1:1.18 | 1;1.18 | 1;1.18 | 1:1.18 | 1;1.18 | 1;1.18 |
| Calcium (%) | 1.07 | 1.08 | 1.09 | 1.08 | 1.09 | 1.10 | 1.12 |

| | | | | | | | |
|---|---|---|---|---|---|---|---|
| Available phosphorus (%) | 0.56 | 0.57 | 0.58 | 0.57 | 0.57 | 0.58 | 0.57 |
| Chemical analyses | | | | | | | |
| DM | 93.52 | 93.47 | 93.71 | 94.11 | 93.17 | 93.40 | 93.98 |
| CP | 22.50 | 22.53 | 22.25 | 21.84 | 22.20 | 22.40 | 22.60 |
| NDF[4] | 8.72 | 12.20 | 15.73 | 19.30 | 11.67 | 14.67 | 17.68 |
| ADF[5] | 4.45 | 6.67 | 9.09 | 11.42 | 6.55 | 8.66 | 10.79 |
| EE | 6.44 | 7.64 | 9.35 | 10.30 | 7.34 | 8.53 | 9.98 |
| Ash | 3.59 | 4.51 | 4.59 | 4.77 | 4.92 | 4.10 | 4.40 |
| NFE | 64.58 | 61.44 | 59.40 | 58.16 | 61.84 | 60.88 | 58.81 |

DM, dry matter; CP, crude protein; EE, ether extract; NFE, nitrogen free extract.

[a] Provided per kg diet: vitamin A 6670 IU; vitamin $D_3$ 1000 IU; vitamin E 23 IU; vitamin $K_3$ 1.33 mg; cobalamin 0.03 mg; Thiamin 0.83 mg; riboflavin 2.0 mg; folic acid 0.33 mg; biotin 0.03 mg; pantothenic acid 3.75 mg; niacin 23.30 mg; pyridoxine 1.33 mg.

[b] Provided per kg diet: Fe 100.0 mg; Mn 110.0 mg; Cu 20.0 mg; Zn 100.0 mg; I 2.0 mg; Se 0.20 mg; Co 0.60 mg.

[c] Butyrated hydroxytoluene (BHT).

[d] Diets were formulated using FeedLIVE software (FeedLIVE 1.52, Thailand).

[1] PKC: Palm kernel cake.

[2] FPKC: Fermented palm kernel cake.

[3] DCP: Dicalcium phosphate.

[4] Neutral detergent fiber.

[5] Acid detergent fibe.

[6] ME: Metabolizable energy.

## 2.3. Nutrient digestibility

Titanium dioxide ($TiO_2$) was added to the feed as an indigestible marker at 0.3% before slaughtering the birds for four days during the end of the starter and finisher phases. Ileal digesta were collected at 3 and 6 weeks of age. The $TiO_2$ was determined based on the method described by Short et al. (1996). Proximate analysis was applied to the feed and digesta to calculate the DM, CP, EE, ash NDF and ADF digestibilities based on the methods mentioned above.

**Table 2.** Ingredient and chemical compositions of broiler finisher diet.

| Ingredient (%) | PKC[1] level (%) | | | | FPKC[2] level (%) | | |
|---|---|---|---|---|---|---|---|
| | 0 | 5 | 10 | 15 | 5 | 10 | 15 |
| Yellow corn | 60.50 | 54.95 | 49.39 | 43.84 | 54.95 | 49.39 | 43.84 |
| Palm oil | 2.60 | 4.20 | 5.77 | 7.32 | 4.20 | 5.77 | 7.32 |
| PKC | 0.00 | 5.00 | 10.00 | 15.00 | 0.00 | 0.00 | 0.00 |
| FPKC | 0.00 | 0.00 | 0.00 | 0.00 | 5.00 | 10.00 | 15.00 |
| Soybean meal 44% CP | 28.16 | 26.80 | 25.70 | 24.60 | 26.80 | 25.70 | 24.60 |
| Fish meal 55% CP | 5.00 | 5.00 | 5.00 | 5.00 | 5.00 | 5.00 | 5.00 |
| DCP[3] 18% | 1.69 | 1.80 | 1.80 | 1.80 | 1.80 | 1.80 | 1.80 |
| Calcium carbonate | 0.30 | 0.37 | 0.37 | 0.37 | 0.37 | 0.37 | 0.37 |
| DL-Methionine | 0.20 | 0.25 | 0.28 | 0.28 | 0.25 | 0.28 | 0.28 |
| L-Lysine | 0.00 | 0.10 | 0.16 | 0.23 | 0.10 | 0.16 | 0.23 |
| L-Threonine | 0.00 | 0.00 | 0.00 | 0.03 | 0.00 | 0.00 | 0.03 |
| Salt | 0.32 | 0.30 | 0.30 | 0.30 | 0.30 | 0.30 | 0.30 |
| Vitamin Premix[a] | 0.50 | 0.50 | 0.50 | 0.50 | 0.50 | 0.50 | 0.50 |
| Mineral Premix[b] | 0.50 | 0.50 | 0.50 | 0.50 | 0.50 | 0.50 | 0.50 |
| Toxin binder | 0.14 | 0.14 | 0.14 | 0.14 | 0.14 | 0.14 | 0.14 |
| Antioxidant[c] | 0.01 | 0.01 | 0.01 | 0.01 | 0.01 | 0.01 | 0.01 |
| Choline chloride | 0.08 | 0.08 | 0.08 | 0.08 | 0.08 | 0.08 | 0.08 |
| Total | 100 | 100 | 100 | 100 | 100 | 100 | 100 |
| Calculated analyses[d] | | | | | | | |
| ME[6] (kcal/kg) | 3000.5 | 3000.2 | 3000.5 | 3000.0 | 3000.2 | 3000.5 | 3000.0 |
| Methionine + Cystine (%) | 0.90 | 0.91 | 0.91 | 0.90 | 0.90 | 0.91 | 0.90 |
| Lysine (%) | 1.10 | 1.15 | 1.18 | 1.21 | 1.15 | 1.18 | 1.21 |
| Arginine (%) | 1.33 | 1.36 | 1.40 | 1.43 | 1.36 | 1.40 | 1.43 |
| Lys: Arg ratio | 1:1.21 | 1:1.18 | 1:1.19 | 1:1.18 | 1:1.18 | 1:1.19 | 1:1.18 |

| | | | | | | | |
|---|---|---|---|---|---|---|---|
| Calcium (%) | 1.01 | 1.09 | 1.08 | 1.14 | 1.07 | 1.05 | 1.14 |
| Available phosphorus (%) | 0.57 | 0.58 | 0.57 | 0.58 | 0.58 | 0.57 | 0.58 |
| Chemical analyses | | | | | | | |
| DM | 90.00 | 90.50 | 91.60 | 91.00 | 91.10 | 91.61 | 90.10 |
| CP | 20.12 | 20.31 | 19.97 | 20.02 | 20.00 | 20.20 | 20.10 |
| NDF[4] | 8.58 | 12.07 | 15.60 | 19.12 | 11.54 | 14.54 | 17.53 |
| ADF[5] | 4.22 | 6.52 | 8.85 | 11.20 | 6.31 | 8.42 | 10.54 |
| EE | 5.70 | 7.27 | 9.67 | 11.73 | 7.86 | 9.66 | 11.53 |
| Ash | 6.29 | 5.19 | 5.71 | 5.10 | 5.21 | 6.02 | 5.27 |
| NFE | 64.76 | 63.27 | 60.23 | 58.47 | 63.86 | 60.12 | 59.05 |

DM, dry matter; CP, crude protein; EE, ether extract; NFE, nitrogen free extract.

[a] Provided per kg diet: vitamin A 6670 IU; vitamin $D_3$ 1000 IU; vitamin E 23 IU; vitamin $K_3$ 1.33 mg; cobalamin 0.03 mg; Thiamin 0.83 mg; riboflavin 2.0 mg; folic acid 0.33 mg; biotin 0.03 mg; pantothenic acid 3.75 mg; niacin 23.30 mg; pyridoxine 1.33 mg.

[b] Provided per kg diet: Fe 100.0 mg; Mn 110.0 mg; Cu 20.0 mg; Zn 100.0 mg; I 2.0 mg; Se 0.20 mg; Co 0.60 mg.

[c] Butyrated hydroxytoluene (BHT).

[d] Diets were formulated using FeedLIVE software (FeedLIVE 1.52, Thailand).

[1] PKC: Palm kernel cake.

[2] FPKC: Fermented palm kernel cake.

[3] DCP: Dicalcium phosphate.

[4] Neutral detergent fiber.

[5] Acid detergent fiber.

[6] ME: Metabolizable energy.

## 2.4. Morphology of small intestines

The procedure for measuring the villi height and crypt depth was according to the method described by Choe et al. (2012). Five samples were taken at three weeks and ten samples at six weeks of age. Segments of approximately 5–6 cm long were removed from three different parts of the small intestine. The portion of duodenum was taken from the middle part of the duodenum loop, and for the jejunum, from midway between the end of the duodenal loop and Meckel's diverticulum, whereas the segment for ileum was taken from midway between Meckel's diverticulum and the ileo-cecal junction. All intestinal segments were flushed with 10% (v/v) neutral buffered formalin solution. The sections were filled out with the neutral buffered formalin, and banded before being kept in the formalin buffer solution for further morphometric analysis. The intestinal segments were excised to approximately 3 mm, and transferred into plastic cassettes in order to measure villus height and crypt depth. The whole excised segments were kept overnight in neutral buffer formalin solution. Then, the intestinal samples were dehydrated in tissue processing machine (Leica ASP 3000, Japan), and embedded in paraffin wax (Leica EG 1160, Japan). The intestinal samples were trimmed at 15 μm. Later, the samples were sectioned at 5 μm, and fixed on slides in hot plate at 60 °C. The slides were stained with hematoxylin and eosin, mounted and examined under microscope.

The morphometric variables being examined were villi height and crypt depth. The villus height was measured from the tip of the villus to the base, whereas the crypt depth was measured as the depth of invagination between adjacent villi. Ten villi and crypts were measured from each slide (Five and ten slides at 3 and 6 weeks of age, respectively) and the average was taken.

## 2.5. Gut microflora count

### 2.5.1. Ileal lactic acid bacteria count

The gut microflora count was carried out based on the method described by Foo et al. (2003) using sterile peptone water; 10% (w/v) of ileal digesta were diluted and left at room temperature for one hour. Then, followed by ten-fold serial dilutions (v/v) using sterile peptone water. To determine the colony forming unit (CFU), LAB 100 μl was pipetted into MRS agar (Lactobacillus-Agar De Man, ROGOSA, and SHAPE) (Merck®, KGaA, Darmstadt). Hence, the plates were spread using a sterile L-shape and incubated in an anaerobic jar at 30 °C for 48 h.

### 2.5.2. Ileal Enterobacteriaceae count

The petri dishes were spread plated in Eosin-methylene blue lactose sucrose agar (Merck®, KGaA, Darmstadt) in order to determine the ENT count, and incubated anaerobically at 37 °C for 24 h. The CFU was expressed as the base of 10 logarithms of CFU per gram digesta, and all samples were repeated five times.

## 2.6. Experimental design and data analysis

The experimental design was based on completely randomized design, and data were analyzed using the General Linear Model procedure of the statistical analysis system (SAS, 2003). Orthogonal polynomial contrasts were used to determine the linear and quadratic effects of dietary increasing proportion of inclusion PKC or FPKC against the control group on nutrient digestibility and gut microflora. Tukey's test was used to compare the means of treatment with probability 5% ($P < 0.05$). The statistical model used was $Y_{ijk} = \mu + T_{ij} + E_{ijk}$ where, $Y_{ijk}$, response variables; $\mu$, the overall mean; $T_{ij}$, the effect of dietary treatment; $E_{ijk}$, the experimental error.

# 3. RESULTS

## 3.1. Nutrient digestibility

The effect of feeding FPKC on the nutrient digestibility of broiler chickens is shown inTable 3. The findings showed that feeding 10% or 15% PKC in the starter and finisher phases led to a significant linear decrease ($P < 0.05$) in the digestibility of dry matter, crude protein, ether extract and nitrogen free extract compared to those birds fed with the negative control. On the other hand, no significant differences ($P > 0.05$) were observed in the dry matter, crude protein, and ether extract digestibility for those fed with 10% or 15% FPKC in comparison with the negative control in two phases, whereas a significant linear and quadratic improvement ($P < 0.05$) in the nutrient digestibility was found for those birds fed with 10% or 15% FPKC compared to those groups of birds fed with PKC.

A significant linear decrease ($P < 0.0049$) at starter phase in NDF digestibility, and both linear ($P < 0.0001$) and quadratic decrease ($P < 0.0088$) at finisher phase in NDF digestibility was observed for broilers fed with 15% PKC compared to those birds fed with the negative control diet. A significant linear and quadratic decrease ($P < 0.0001$) was also observed in ADF digestibility at finisher phase.

**Table 3.** Effect of fermented palm kernel cake on ileal nutrient digestibility on broiler chickens.

| Nutrient (%) | PKC[1] level | | | | FPKC[2] level | | | | Contrast, P-values | |
|---|---|---|---|---|---|---|---|---|---|---|
| | 0% | 5% | 10% | 15% | 5% | 10% | 15% | SEM[5] | Linear[3] | Quadratic[4] |
| 21 days | | | | | | | | | | |
| DM | $61.32^{a}$ | $59.26^{b,c}$ | $58.38^{c}$ | $57.89^{c}$ | $60.63^{a,b}$ | $60.59^{a,b}$ | $60.77^{a,b}$ | 0.38 | <0.0001 | 0.0006 |
| CP | $80.87^{a}$ | $79.89^{a}$ | $76.21^{b,c}$ | $75.01^{c}$ | $79.93^{a}$ | $78.97^{a,b}$ | $79.73^{a}$ | 0.65 | <0.0001 | 0.0024 |
| NDF | $35.81^{a}$ | $36.10^{a}$ | $35.03^{a,b}$ | $33.85^{b}$ | $36.06^{a}$ | $35.38^{a,b}$ | $35.10^{a,b}$ | 0.35 | 0.0049 | 0.2194 |
| ADF | 20.50 | 20.00 | 19.88 | 19.75 | 20.38 | 20.50 | 20.75 | 0.48 | 0.7199 | 0.7489 |
| EE | $54.95^{a}$ | $54.11^{a,b}$ | $53.20^{b,c}$ | $52.50^{c}$ | $54.86^{a}$ | $54.52^{a,b}$ | $53.96^{a,b,c}$ | 0.36 | 0.0006 | 0.0910 |
| NFE | $63.93^{a}$ | $62.80^{a,b}$ | $58.94^{c}$ | $58.51^{c}$ | $63.59^{a,b}$ | $61.92^{b}$ | $61.79^{b}$ | 0.40 | <0.0001 | 0.0128 |
| Ash | 34.74 | 34.42 | 33.88 | 33.60 | 34.94 | 34.68 | 34.37 | 0.40 | 0.2387 | 0.4863 |
| 42 days | | | | | | | | | | |
| DM | $67.11^{a}$ | $64.59^{b}$ | $62.08^{c}$ | $61.38^{c}$ | $65.95^{a,b}$ | $66.22^{a,b}$ | $68.30^{a}$ | 0.51 | <0.0001 | <0.0001 |
| CP | $77.87^{a}$ | $77.08^{a}$ | $65.93^{b}$ | $61.93^{c}$ | $76.01^{a}$ | $76.20^{a}$ | $75.60^{a}$ | 0.61 | <0.0001 | 0.0001 |
| NDF | $48.30^{a}$ | $47.26^{a,b}$ | $45.20^{c,d}$ | $44.43^{d}$ | $46.68^{a,b,c}$ | $45.76^{b,c,d}$ | $46.52^{a,b,c}$ | 0.44 | <0.0001 | 0.0088 |
| ADF | $25.88^{a}$ | $25.13^{a,b}$ | $23.25^{c,d}$ | $22.25^{d}$ | $24.88^{a,b,c}$ | $24.12^{b,c}$ | $24.20^{a,b,c}$ | 0.54 | <0.0001 | <0.0001 |
| EE | $69.22^{a}$ | $68.00^{a,b}$ | $65.90^{b}$ | $63.63^{c}$ | $68.77^{a}$ | $67.81^{a,b}$ | $68.42^{a}$ | 0.48 | <0.0001 | 0.0029 |
| NFE | $67.83^{a}$ | $66.96^{a}$ | $63.11^{b}$ | $62.77^{b}$ | $67.70^{a}$ | $66.55^{a}$ | $66.08^{a}$ | 0.49 | <0.0001 | 0.0220 |
| Ash | $55.48^{a}$ | $54.18^{a,b}$ | $53.04^{a,b}$ | $51.83^{b}$ | $54.22^{a,b}$ | $55.34^{a}$ | $53.96^{a,b}$ | 0.71 | 0.0216 | 0.0677 |

DM; dry matter, CP; crude protein, NDF; neutral detergent fiber, ADF; acid detergent fiber, EE; ether extract, NFE; nitrogen-free extract.

n = 5 at 3 weeks and 10 at 6 weeks of age.

[a,b,c,d] Means ± SEM. Means with different superscripts in the same row differ significantly ($P < 0.05$).

[1] PKC: Palm kernel cake.

[2] FPKC: Fermented palm kernel cake.

[3] Linear and quadratic responses, respectively, to PKC or FPKC inclusion levels.

[4] Linear and quadratic responses, respectively, to PKC or FPKC inclusion levels.

[5] Pooled standard error.

The digestibility of ash was linearly lower ($P = 0.0216$) than the negative control for chickens fed with 15% PKC. However, the digestibility of ash did not significantly ($P > 0.05$) differ for those groups of birds fed with 10% or 15% FPKC as compared to the negative control at the sixth week of age.

## 3.2. Morphology of small intestines

The effect of feeding FPKC on villus height and crypt depth in broiler chickens is shown inTable 4. The data obtained did not show any significant differences ($P > 0.05$) among the dietary treatments in villus height and crypt depth for three segments of the small intestine – duodenum, jejunum and ileum – in the starter or finisher phase.

**Table 4.** Effect of fermented palm kernel cake on villus height and crypt depth in broiler chickens.

| Dietary treatment | Villi height (µm) | | | Crypt depth (µm) | | |
|---|---|---|---|---|---|---|
| | Duodenum | Jejunum | Ileum | Duodenum | Jejunum | Ileum |
| | | 21 days | | | | |
| 0% PKC[1] | 1025.10 | 548.12 | 333.70 | 187.23 | 130.33 | 103.25 |
| 5% PKC | 1031.21 | 568.93 | 383.63 | 196.06 | 144.83 | 101.95 |
| 10% PKC | 1105.13 | 549.99 | 386.51 | 184.60 | 137.26 | 106.71 |
| 15% PKC | 1109.33 | 550.32 | 358.70 | 188.53 | 131.90 | 109.22 |
| 5% FPKC[2] | 1123.28 | 529.04 | 365.29 | 194.93 | 133.00 | 112.35 |
| 10% FPKC | 1107.53 | 517.05 | 349.67 | 204.21 | 141.37 | 110.94 |
| 15% FPKC | 1127.36 | 545.62 | 391.42 | 191.99 | 137.20 | 106.11 |
| SEM[3] | 27.31 | 21.40 | 15.10 | 6.60 | 8.06 | 2.82 |
| Contrast, P-values | | | | | | |
| Linear[3] | 0.0757 | 0.7228 | 0.0788 | 0.052 | 0.8599 | 0.0998 |
| Quadratic[4] | 0.0687 | 0.4934 | 0.5506 | 0.0622 | 0.9754 | 0.518 |
| | | 42 days | | | | |
| 0% PKC | 1473.14 | 735.56 | 614.34 | 227.67 | 213.26 | 197.92 |
| 5% PKC | 1458.61 | 736.38 | 609.41 | 229.80 | 219.06 | 201.04 |
| 10% PKC | 1440.86 | 730.43 | 601.30 | 226.15 | 212.13 | 178.67 |
| 15% PKC | 1439.65 | 734.91 | 614.37 | 224.68 | 202.88 | 194.18 |
| 5% FPKC | 1471.21 | 732.08 | 613.14 | 233.26 | 217.13 | 208.23 |
| 10% FPKC | 1480.77 | 731.34 | 607.71 | 231.04 | 217.71 | 200.81 |
| 15% FPKC | 1482.87 | 733.50 | 669.30 | 234.35 | 216.01 | 211.75 |
| SEM | 30.50 | 15.50 | 17.30 | 5.52 | 5.43 | 9.54 |
| Contrast, P-values | | | | | | |
| Linear[3] | 0.8782 | 1.000 | 0.1233 | 0.8586 | 0.4177 | 0.2840 |
| Quadratic[4] | 0.4200 | 0.7939 | 0.0656 | 0.4706 | 0.6052 | 0.1759 |

n = 5 at 3 weeks and 10 at 6 weeks of age.

[1] PKC: Palm kernel cake.

[2] FPKC: Fermented palm kernel cake.

[3] Linear and quadratic responses, respectively, to PKC or FPKC inclusion levels.

[4] Linear and quadratic responses, respectively, to PKC or FPKC inclusion levels.

[a] Pooled standard error.

## 3.3. Gut microflora

The effect of feeding FPKC on ileal LAB and ENT counts in broiler chickens at three and six weeks of age is presented in Table 5. The results showed that the ENT counts were significantly lower (linear and quadratic, $P < 0.0001$) for the birds fed with 15% FPKC compared to the negative control group or those birds fed with different levels of PKC. No significant difference was observed (linear, $P = 0.0599$; quadratic, $P = 0.1899$) between the dietary treatments in the LAB count at three weeks of age. In contrast, LAB count was significantly higher (linear, $P = 0.0266$; quadratic, $P = 0.0007$) for birds fed with 15% FPKC than the negative control group at six weeks of age, whereas, no significant difference was observed (linear, $P = 0.8077$; quadratic, $P = 0.7240$) between the dietary treatments in ENT counts at the sixth week of age.

# 4. DISCUSSION

## 4.1. Nutrient digestibility

The nutrient digestibility declined in broiler chickens fed with 10% or 15% PKC for both phases. The reduction of apparent ileal digestibility of nutrients could be attributed to the presence of the high levels of NDF, ADF, insoluble fibers and NSPs in the PKC (Sulabo et al., 2013 and Son et al., 2014). Indeed, the viscosity would increase in the small intestines and speed up the passage rate of ingesta (Dégen et al., 2007), and the apparent ileal digestibility of the nutrients would decrease. These findings are consistent with Aya et al. (2013), who reported that the inclusion of 10% PKC in the broiler starter diet led to a significant decrease ($P < 0.05$) in nutrient digestibility. The findings are also in agreement with Lawal et al. (2010), who referred to the reduction of nutrient digestibility for broiler chickens fed with 10% PKC compared to those chickens fed with 10% PKC treated with cellulolytic crude enzymes that hydrolyze fibers. Similar findings were obtained by Mok et al. (2013), who showed that dry matter, organic matter and energy digestibility in growing pigs fed with 10% PKC were significantly ($P < 0.05$) lower than those growing pigs fed with 10% PKC supplemented with β-mannanase.

The results are also in agreement with Brenes et al. (2002), who mentioned that feeding young chicks with a diet containing high levels of fibers (Lupin) had a significantly ($P < 0.05$) adverse effect on the apparent digestibility of protein, whereas a diet supplemented with enzymes (0.1% Novozyme®) led to improved crude protein digestibility. The supplementation of enzyme to a lupin-based diet enhances the apparent ileal digestibility of NSPs in broiler chickens (Brenes et al., 2002).

The findings are in agreement with Fasuyi et al. (2014), who reported that apparent protein digestibility was significantly ($P < 0.05$) lower in broiler chickens fed with 10% PKC compared to the chickens fed with the negative control diet. On the other hand, the apparent protein digestibility was improved for those chickens fed with 10% PKC supplemented with enzymes (cellulase, glucanase, and xylanase). It has been reported that *P. polymyxa* ATCC 842 was able to produce high cellulase, xylanase, and mannanase activities on different substrates ( Alshelmani et al., 2013).

Similar findings were obtained by Selle et al. (2009), who claimed that a diet supplemented with xylanase significantly ($P < 0.05$) enhanced the ileal digestibility of protein for broiler chickens fed with a wheat-based diet compared to those chickens fed with the control diet. It has been reported that apparent ileal digestibility of crude protein can be reduced by 0.3%–0.8% per every percentage point in NDF increase (Sulabo et al., 2013). In addition, the adverse effect of soluble fibers (NSPs) is greater than the insoluble fibers in terms of nutrient digestibility (Dégen et al., 2007 and Sulabo et al., 2013). Based on our previous work, the reducing sugars have been significantly increased ($P < 0.05$) as a result of NSPs degradation by *P. polymyxa* ATCC 842 ( Alshelmani et al., 2014). This can interpret the improvement in nutrient digestibility of FPKC.

**Table 5.** Effect of fermented palm kernel cake on ileal lactic acid bacteria (LAB) and *Enterobacteriaceae* (ENT) count in broiler chickens.

| Bacterial count, $log_{10}$ CFU/g | | PKC level | | | | FPKC level | | | | Contrast, P-values | |
|---|---|---|---|---|---|---|---|---|---|---|---|
| | | 0% | 5% | 10% | 15% | 5% | 10% | 15% | SEM[3] | Linear[1] | Quadratic[2] |
| 21 days | | | | | | | | | | | |
| | LAB count | 5.20 | 5.25 | 5.27 | 5.17 | 5.28 | 5.24 | 5.30 | 0.03 | 0.0599 | 0.1899 |
| | ENT count | 4.19$^{a,b}$ | 4.24$^{a,b}$ | 4.27$^{a}$ | 4.22$^{a,b}$ | 4.20$^{a,b}$ | 4.13$^{b,c}$ | 4.03$^{c}$ | 0.02 | <0.0001 | <0.0001 |
| 42 days | | | | | | | | | | | |
| | LAB count | 5.17$^{b}$ | 5.26$^{a,b}$ | 5.27$^{a,b}$ | 5.44$^{a,b}$ | 5.43$^{a,b}$ | 5.44$^{a,b}$ | 5.56$^{a}$ | 0.07 | 0.0266 | 0.0007 |
| | ENT count | 4.38 | 4.67 | 4.60 | 4.42 | 4.26 | 4.31 | 4.45 | 0.22 | 0.8077 | 0.7240 |

n = 5 at 3 weeks and 10 at 6 weeks of age.

[a,b,c] Means ± SEM. Means with different superscripts in the same row are differ significantly ($P < 0.05$).

[1] Linear and quadratic responses, respectively, to PKC or FPKC inclusion levels.

[2] Linear and quadratic responses, respectively, to PKC or FPKC inclusion levels.

[3] Pooled standard error.

The reduction of nutrient digestibility could be attributed not only to the presence of insoluble fibers, but also to the high content of soluble fibers, such as NSPs. These molecules can increase the speed passage rate of ingesta, which may result in decreased nutrient absorption. Therefore, high levels of dietary fiber lead to a reduction in the digestibility of energy, starch, protein and lipids in monogastric animals (Montagne et al., 2003 and Walugembe et al., 2014). In the current experiment, it seems that the full genetic potential of the commercial meat-type chickens may not be able to be expressed because of environmental constraints (Awad et al., 2014).

## 4.2. Morphology of small intestines

It has been reported that feed containing a high content of dietary fiber can decrease the surface area, width and height of intestinal villi (Moharrery and Mohammadpour, 2005;Kalmendal et al., 2011 and Carabaño et al., 2013). The effect of dietary fiber on villi height is variable and depends on the physio-chemical characteristics of the fibers, levels, duration of the dietary fiber in the diet, animal species, age (Montagne et al., 2003) and health status of the birds (Mateos et al., 2012).

In the current study, no significant differences were found for the villi height and crypt depth in the small intestine. However, these findings are consistent with Zulkifli et al. (2009), who claimed that no significant difference ($P > 0.05$) was observed in the intestinal villi height for broiler chickens fed 25% PKC in the diets.

## 4.3. Gut microflora

The important role of gastrointestinal microflora in the health of the animals has been investigated. LAB are considered to be the normal microflora of the GIT. LAB are known for their ability to produce a wide variety of anti-bacterial substances, such as acetate, lactate, propionate, ethanol and bacteriocins (Loh et al., 2009). These compounds are able to control and inhibit the growth of intestinal pathogenic bacteria (Foo et al., 2003,Nguyen et al., 2010, Choe et al., 2012 and Loh et al., 2014). No significant difference in the results was found among the dietary treatments in the LAB count at three weeks of age, whereas the LAB count significantly ($P < 0.05$) increased in the birds fed with 15% FPKC compared to the negative control in the finisher phase (six weeks of age). These findings are inconsistent with Zulkifli et al. (2009), who reported that the LAB count increased significantly ($P < 0.05$) in the ileum and caecum for finisher broilers fed with 25% PKC. On the other hand, these findings are consistent with the results of Gao et al. (2008), who claimed that there were no significant differences in the LAB count for broilers fed with a wheat-based diet in the starter phase.

The colonization of LAB in the intestinal tract may be influenced by the composition of the diet (Wenk, 2001). It was reported by Zulkifli et al. (2009) that the presence of 78% NSPs in PKC might have a profound effect on gut microflora. Wenk (2001), and Chen et al. (2014) reported that long-term feeding of high ratios of dietary fibers could alter the physiological and anatomical characteristics of the GIT of pigs. The soluble dietary fibers may increase ileal and post-ileal microbial fermentation (Wenk, 2001).

The ileum ENT counts were significantly ($P < 0.05$) decreased at 3 weeks of age, while the LAB count increased at 6 weeks of age in the current study. These findings are in agreement with Zulkifli et al. (2009), who mentioned that an increase in the LAB count can depress the growth of enteropathogens in the gut. The findings are also consistent withLoh et al. (2009), and Choe et al. (2012) who reported that there is a negative correlation between LAB and ENT, in that an increased LAB population may reduce the ENT count.

# 5. CONCLUSION

PKC fermented (FPKC) by *P. polymyxa* ATCC 842 can be offered up to 15% in broiler rations without causing any adverse effects on nutrient digestibility (except in nitrogen free extract at 3 weeks of age). The intestinal villi height and crypt depth were not affected by the inclusion of FPKC in the diet, whereas the LAB count was increased in broilers fed with 15% FPKC. In this way, the inclusion of 15% FPKC can replace up to 30% of yellow corn in the birds' diets, which can be reflected in the cost saving of feed in the poultry industry.

# ACKNOWLEDGMENT

This research project was supported by Long-Term Research Grant Scheme (LRGS) provided by the Ministry of Higher Education, Malaysia.

# REFERENCES

1. AOAC, 1990. Association of Official Analytical Chemists. Official Method of Analysis, 15th ed. Washington D.C., USA.

2. Ahmed, A., Zulkifli, I., Farjam, A.S., Abdullah, N., Liang, J.B., Awad, E.A., 2014. Effect of solid state fermentation on nutrient content and ileal amino acids digestibility of canola meal in broiler chickens. Italian Journal of Animal Science 13, 410–414.

3. Alimon, A.R., 2004. The nutritive value of palm kernel cake for animal feed. Palm Oil Dev. 40, 12–14.

4. Alshelmani, M.I., Loh, T.C., Foo, H.L., Lau, W.H., Sazili, A.Q., 2013. Characterization of cellulolytic bacterial cultures grown in different substrates. Sci. World J. 2013, 6.

5. Alshelmani, M.I., Loh, T.C., Foo, H.L., Lau, W.H., Sazili, A.Q., 2014. Biodegradation of palm kernel cake by cellulolytic and hemicellulolytic bacterial cultures through solid state fermentation. Sci. World J. 2014, 8.

6. Awad, E.A., Fadlullah, M., Zulkifli, I., Farjam, A.S., Chwen, L.T., 2014. Amino acids fortification of low-protein diet for broilers under tropical climate: ideal essential amino acids profile. Ital. J. Anim. Sci. 13, 270–274.

7. Aya, V., Ayanwale, B., Ijaiya, A., Aremu, A., 2013. Performance and nutrient digestibility in broiler chicks as influenced by multienzyme addition to starter diets containing palm kernel meal. Biotechnol. Anim. Husb. 29, 93–104.

8. Brenes, A., Marquardt, R.R., Guenter, W., Viveros, A., 2002. Effect of enzyme addition on the performance and gastrointestinal tract size of chicks fed lupin seed and their fractions. Poult. Sci. 81, 670–678.

9. Carabano, ˜ R., Badiola, I., Chamorro, S., García, J., García-Ruiz, A.I., García-Rebollar, P., Gómez-Conde, M.S., Gutiérrez, I., Nicodemus, N., Villamide, M.J., de Blas, J.C., 2013. New trends in rabbit feeding: influence of nutrition on intestinal health: a review. Span. J. Agric. Res. 6, 15–25.

10. Chen, H., Mao, X., Che, L., Yu, B., He, J., Yu, J., Han, G., Huang, Z., Zheng, P., Chen, D., 2014. Impact of fiber types on gut microbiota, gut environment and gut function in fattening pigs. Anim. Feed Sci. Technol. 195, 101–111.

11. Choe, D.W., Loh, T.C., Foo, H.L., Hair-Bejo, M., Awis, Q.S., 2012. Egg production, faecal pH and microbial population, small intestine morphology, and plasma and yolk cholesterol in laying hens given liquid metabolites produced by Lactobacillus plantarum strains. Br. Poult. Sci. 53, 106–115.

12. Dégen, L., Halas, V., Babinszky, L., 2007. Effect of dietary fibre on protein and fat digestibility and its consequences on diet formulation for growing and fattening pigs: a review. Acta Agric. Scand. Sect. A—Anim. Sci. 57, 1–9.

13. Fasuyi, A., Abiodun, S., Akomolafe, O., 2014. Bioconversion and enzymes fortification of palm kernel meal as protein supplement in broiler rations. Am. J. Exp. Agric. 4, 767–784.

14. Foo, H.L., Loh, T.C., Lim, H.A., Shukriyah, M.H., Kufli, C.N., Law, F.L., 2003. Effects of fermented fruits on growth performance, shedding of enterobacteriaceae and Lactic acid bacteria and plasma cholesterol in rats. Pak. J. Nutr. 2, 228–233.

15. Francesch, M., Brufau, J., 2004. Nutritional factors affecting excreta/litter moisture and quality. World' Poult. Sci. J. 60, 64–75.

16. Gao, F., Jiang, Y., Zhou, G.H., Han, Z.K., 2008. The effects of xylanase supplementation on performance, characteristics of the gastrointestinal tract, blood parameters and gut microflora in broilers fed on wheat-based diets. Anim. Feed Sci. Technol. 142, 173–184.

17. Goering, H.K., Van Soest, P.J., 1970. Forage fiber analysis (apparatus, reagents, prosedures and some applications). USDA. Agricultural Handbook. No 379. Kalmendal, R., Elwinger, K., Holm, L., Tauson, R., 2011. High-fibre sunflower cake affects small intestinal digestion and health in broiler chickens. Br. Poult. Sci. 52, 86–96.

18. Lawal, T.E., Iyayi, E.A., Adeniyi, B.A., Adaramoye, O.A., 2010. Biodegradation of palm kernel cake with multienzyme complexes from fungi and its feeding value for broiler. Int. J. Poult. Sci. 9, 695–701.

19. Loh, T.C., Chong, S.W., Foo, H.L., Law, F.L., 2009. Effects on growth performance, faecal microflora and plasma cholesterol after supplementation of spray-dried metabolite to postweaning rats. Czech J. Anim. Sci. 54, 10–16.

20. Loh, T.C., Choe, D.W., Foo, H.L., Sazili, A.Q., Bejo, M.H., 2014. Effects of feeding different postbiotic metabolite combinations produced by Lactobacillus plantarum strains on egg quality and production performance, faecal parameters and plasma cholesterol in laying hens. BMC Vet. Res. 10, 149.

21. Marini, A.M., Daud, M.J., Noraini, S., Jameah, H., Azahan, E.A.E., 2005. Performance of locally isolated microorganism in degrading palm kernel cake (PKC) fiber and improving the nutritional value of fermented PKC. J. Trop. Agric. Food Sci. 33, 311–319.

22. Mateos, G.G., Jiménez-Moreno, E., Serrano, M.P., Lázaro, R.P., 2012. Poultry response to high levels of dietary fiber sources varying in physical and chemical characteristics. J. Appl. Poult. Res. 21, 156–174.

23. McDonald, P., Edwards, R.A., Greenhalgh, J.F.D., 1995. Animal Nutrition. Longman Singapore Publishers Pte Ltd.

24. Moharrery, A., Mohammadpour, A., 2005. Effect of diets containing different qualities of barley on growth performance and serum amylase and intestinal villus morphology. Int. J. Poult. Sci. 4, 549–556.

25. Mok, C.H., Lee, J.H., Kim, B.G., 2013. Effects of exogenous phytase and -mannanase on ileal and total tract digestibility of energy and nutrient in palm kernel expeller-containing diets fed to growing pigs. Anim. Feed Sci. Technol. 186, 209–213.

26. Montagne, L., Pluske, J.R., Hampson, D.J., 2003. A review of interactions between dietary fibre and the intestinal mucosa, and their consequences on digestive health in young non-ruminant animals. Anim. Feed Sci. Technol. 108, 95–117.

27. Muangkeow, N., Chinajariyawong, C., 2009. Determination of true amino acid digestibility and metabolizable energy in fermented palm kernel meal with Aspergillus wentii TISTR 3075 for chickens. Walailak J. Sci. Technol. 6, 231–241.

28. NRC, 1994. Nutrient Requirements of Poultry, 9th rev. ed. National Academy Press, Washington, DC.

29. Nguyen, T.T., Loh, T.C., Foo, H.L., Bejo, M.H., Kasim, A., 2010. Inhibitory activity of metabolites produced by strains of Lactobacillus plantarum isolated from Malaysian fermented food. Int. J. Probiotics Prebiotics 5, 37–44.

30. O'Mara Mulligan, F.P., Cronin, F.J., Rath, E.J.M., Caffrey, P.J., 1999. The nutritive value of palm kernel meal measured in vivo and using rumen fluid and enzymatic techniques. Livest. Prod. Sci. 60, 305–316.

31. SAS, 2003. Statistical Analytical System. SAS Institute Inc Cary, NC, USA.

32. Selle, P.H., Ravindran, V., Partridge, G.G., 2009. Beneficial effects of xylanase and/or phytase inclusions on ileal amino acid digestibility, energy utilisation, mineral retention and growth performance in wheat-based broiler diets. Anim. Feed Sci. Technol. 153, 303–313.

33. Short, F., Gorton, P., Wiseman, J., Boorman, K., 1996. Determination of titanium dioxide added as an inert marker in chicken digestibility studies. Anim. Feed Sci. Technol. 59, 215–221.

34. Son, A.R., Hyun, Y., Htoo, J.K., Kim, B.G., 2014. Amino acid digestibility in copra expellers and palm kernel expellers by growing pigs. Anim. Feed Sci. Technol. 187, 91–97.

35. Sulabo, R.C., Ju, W.S., Stein, H.H., 2013. Amino acid digestibility and concentration of digestible and metabolizable energy in copra meal, palm kernel expellers, and palm kernel meal fed to growing pigs. J. Anim. Sci. 91, 1391–1399.

36. Sundu, B., Dingle, J., 2002. Use of enzymes to improve the nutritional value of palm kernel meal and copra meal , Proceeding of the Queensland Poultry Science Symposium, Australia, pp. 1–15.

37. Walugembe, M., Rothschild, M.F., Persia, M.E., 2014. Effects of high fiber ingredients on the performance, metabolizable energy and fiber digestibility of broiler and layer chicks. Anim. Feed Sci. Technol. 188, 46–52.

38. Wenk, C., 2001. The role of dietary fibre in the digestive physiology of the pig. Anim. Feed Sci. Technol. 90, 21–33.

39. Zulkifli, I., Rahayu, H.I., Alimon, A., Vidyadaran, M., Babjee, S., 2009. Gut microflora and intestinal morphology of commercial broiler chickens and red jungle fowl fed diets containing palm kernel meal. Archiv für Geflügelkunde 73, 49–55.

# CHAPTER 5

# Impact of mycotoxin on immune response and consequences for pig health

**Alix Pierron[a, b, c], Imourana Alassane-Kpembi[a, b], Isabelle P. Oswald[a, b]**

[a] INRA, UMR 1331, ToxAlim Research Centre in Food Toxicology, BP93173, Toulouse Cedex 03 31027, France

[b] Universit_e de Toulouse, INP, UMR 1331, ToxAlim, BP93173, Toulouse Cedex 03 31027, France

[c] BIOMIN Research Center, Technopark 1, Tulln 3430, Austria

## ABSTRACT

Mycotoxins are fungal secondary metabolites detected in many agricultural commodities, especially cereals. Due to their high consumption of cereals, pigs are exposed to these toxins. In the European Union, regulations and/or recommendations exist in pig feed for aflatoxins, ochratoxin A, fumonisins, zearalenone, and trichothecenes, deoxynivalenol and T-2 toxin. These mycotoxins have different toxic effects, but they all target the immune system. They have immunostimulatory or immunosuppressive effects depending on the toxin, the concentration and the parameter investigated. The immune system is primarily responsible for defense against invading organisms. The consequences of the ingestion of mycotoxin-contaminated feed are an increased susceptibility to infectious diseases, a reactivation of chronic infection and a decreased vaccine efficacy. In this review we summarized the data available on the effect of mycotoxins on the immune system and the consequences for pig health.

## KEYWORDS

Pig; Mycotoxins; Feed contamination; Susceptibility to disease; Immunity; Vaccine efficacy

## 1. INTRODUCTION

Mycotoxins are toxic secondary metabolites produced by various molds, such as*Aspergillus*, *Penicillium* and *Fusarium*, which may contaminate food and feed at all stages of the food/feed chain. Despite the improvement of good agricultural and manufacturing practices, mycotoxin contamination cannot be avoided and contaminants are virtually ubiquitous at some concentrations in the average human and animal diets ( Bryden, 2012). A recent study performed on 1100 samples collected worldwide showed that about 70% of samples tested was contaminated (Streit et al., 2013). This result was confirmed on a smaller study realized on 83 feed ingredients sampled in China (Guan et al., 2011a).

The biological reactions following ingestion of mycotoxins vary from acute, overt diseases with high morbidity and mortality to chronic, insidious disorders with reduced animal productivity. Different mycotoxins target different organs, inducing various toxic effects. At high doses, mycotoxins exposure leads to general cytotoxicity, often related to macromolecule synthesis inhibition (Maresca and Fantini, 2010). Mycotoxins induce primary biochemical lesions and impact on early cellular functions/events in the cascade of events leading to toxic cell injury or cellular deregulation (Bryden, 2012). At low doses, mycotoxins affect the functions of various tissues and organs, such as the gastrointestinal tract, liver or kidney tissues, as well as the nervous, reproductive and immune systems. Some mycotoxins also have genotoxic, carcinogenic and teratogenic effects (Maresca and Fantini, 2010).

Mycotoxins contamination levels in pig feedstuffs are usually not high enough to cause an overt disease but may result in economical loss through changes in growth, production and immunosuppression (Bryden, 2012, Oswald et al., 2005 and Wild and Gong, 2010).

Pigs are very sensitive to mycotoxins. Due to their high consumption of cereals, they are exposed to these toxins and to a chronical contamination. In Europe, regulation and/or recommendations exist for 6 mycotoxins that may be present in pig feed: aflatoxins (AF), ochratoxin A (OTA), fumonisins (FB), zearalenone (ZEN) and trichothecenes (principally deoxynivalenol [DON], T-2 and HT-2 toxins) (Bennett and Klich, 2003).

This review summarizes the main effects induced by mycotoxins present in pig feed on immunity and determines the consequences of this immunomodulation in terms of susceptibility to infectious diseases, reactivation of chronic infection and vaccine efficacy.

# 2. EFFECT OF MAJOR MYCOTOXINS ON THE PIG IMMUNE RESPONSE

## 2.1. Aflatoxins

Aflatoxins are hepatotoxic and carcinogenic; they also display immunotoxic properties. These toxins impair both the innate and the acquired immune responses (Meissonnier et al., 2006 and Weaver et al., 2013). The dysregulation of the antigen-presenting capacity of dentritic cells, which is starting from aflatoxin B1 (AFB1) low dose exposure, is deemed to be the mechanism by which the mycotoxin impairs cell-mediated immunity (Mehrzad et al., 2014). An exposition to AF increases the T-cell proliferation-inducing capacity of porcine monocyte-derived dentritic cells, thus enhances presenting capacity of cells (Mehrzad et al., 2015).

An alteration of the inflammatory response has been reported in pigs exposed to AF (Chaytor et al., 2011). A reduced synthesis of pro-inflammatory cytokines and an increase of anti-inflammatory ones was also demonstrated in weanling piglets fed for 4 weeks with low doses of AF (Marin et al., 2002). In utero exposure of piglets to this mycotoxin (through exposition of sows), the functional capacities of both macrophages and neutrophils were altered (Silvotti et al., 1997).

Experimentally, in a pig model vaccinated with a model antigen, which was ovalbumin (OVA), AFB1 exposure had no major effect on humoral immunity with unchanged plasma concentrations of total immunoglobulin A (IgA), IgG and IgM and the specific anti-OVA IgG. In these animals, the toxin exposure did not impair the mitogenic response of lymphocytes but delayed and decreased the OVA-specific proliferation, suggesting an impaired lymphocyte activation in pigs exposed to AFB1 (Meissonnier et al., 2008b). Similarly, in pigs vaccinated with *Mycoplasma*, the exposure to lower levels of AFB1 did not modulate the antigen-specific and total antibody response ( Marin et al., 2002). Developing piglets are very susceptible to this mycotoxin. Indeed, after sows exposure to AF, the global piglets lympho-proliferative response upon mitogenic stimulation is reduced (Silvotti et al., 1997).

## 2.2. Trichothecenes

Type B trichothecenes, including DON, have the capacity to up- and down-regulate immune functions by disrupting intracellular signaling within leukocytes (Pestka, 2010). Depending on the dose, frequency and duration of exposure, DON will have either an immunostimulatory or immunosuppressing effect (Pestka et al., 2004). Deoxynivalenol is able to induce an inflammatory response by acting on the ribosome, inducing a Ribotoxic stress which actives the MAPK pathway, eliciting expression of inflammation-related genes as pro-inflammatory cytokines (Pestka et al., 2004 and Pestka, 2010).

In mice, this toxin induced a pronounced elevation in serum IgA (Pestka et al., 2004). In pigs, a similar increase of IgA in the serum of animals receiving DON contaminated feed has been observed (Drochner et al., 2004, Pinton et al., 2008 and Swamy et al., 2003). In animals immunized with OVA, the specific immune response was investigated during a DON exposure inducing no feed refusal or reduced body weight gain. Ingestion of DON increased the plasma concentration of total and anti-OVA IgA titers. Deoxynivalenol did not modulate lymphocytes proliferation after mitogenic stimulation, but the toxin had a biphasic effect on the OVA-specific lymphocyte proliferation: An up-regulation in the days after OVA immunization but a down-regulation in the weeks following (Pinton et al., 2008).

Another study on pigs immunized with OVA showed an increase of anti-OVA IgG titers, after 42 days of exposure to a DON contaminated diet. Simultaneously, the expressions of chemokines involved in inflammatory reactions (interleukin-8 (IL-8), chemokine (C-X-C motif) ligand 20 (CXCL20), interferon-γ (IFN-γ)) were up-regulated. Deoxynivalenol also up-regulated the gene of major antioxidant glutathione peroxidase 2 (GPX-2) and down-regulated expression of genes encoding enzymatic antioxidants including GPX-3, GPX-4 and superoxide dismutase 3 (SOD-3), involved in oxidative stress (Lessard et al., 2015).

Type A trichothecenes such as T-2 toxin are cytotoxic molecules and potent protein inhibitors. In pigs immunized with OVA, sub-clinical doses of T-2 toxin induced an early and transient increase of total IgA plasma concentration but a decrease in the anti-OVA IgG titer (Meissonnier et al., 2008a). For higher doses of exposure, T-2 toxin had been previously shown to decrease both the mitogenic and the antigen-specific lymphocytes proliferation following a horse globulin immunization (Rafai et al., 1995).

## 2.3. Fumonisins

Fumonisins induce various toxic effects depending on the animal species, and there is evidence for the carcinogenicity of these toxins (Stockmann-Juvala and Savolainen, 2008). In *in vitro* and *in vivo* experiments, fumonisin B1 (FB1) modifies the Th1/Th2 (T-helper 1/T-helper 2) cytokine balance in pigs similar to an impaired humoral response (Marin et al., 2006 and Taranu et al., 2005). With pigs vaccinated against *Mycoplasma* and exposed to FB1 (8 mg/kg feed for 4 weeks), a sex-related difference in the specific immune response has also been observed. In male pigs but not for female ones, exposure to the toxin reduced the vaccine-specific antibody titer ( Marin et al., 2006). However, ingestion of contaminated feed had no effect on the serum concentrations of total IgG, IgA, and IgM.

Studies have also demonstrated that FB1 influences the inflammatory response. For example, incubation of swine alveolar macrophages with FB1 led to a significant reduction of the number of viable cells and cell death by apoptosis (Liu et al., 2002). An*in vivo* experiment on pigs exposed to FB (6 mg/kg feed for 5 weeks) showed a decrease of IL-1β and IL-6 genes expression in spleen tissue ( Grenier et al., 2011).

Fumonisin B1 also impairs on the maturation of antigen presenting cells *in vivo* by reducing the intestinal expression of IL-12p40 and decreasing the upregulation of major histocompatibility complex class II molecule (MHC-II) with a reduction of T cell stimulatory capacity upon stimulation ( Devriendt et al., 2009).

## 2.4. Ochratoxin A

Ochratoxin A is mainly toxic for kidney and liver. Gilts fed OTA-contaminated had reduced cutaneous basophil hypersensitivity response to phytohemagglutinin, reduced delayed hypersensitivity to tuberculin, decreased stimulation index for lymphoblastogenesis, decreased interleukin-2 production when lymphocytes were stimulated with concanavalin A, and decreased number and phagocytic activity of macrophages. Ochratoxin A was shown to be toxic on purified lymphocytes of pigs with an half maximal inhibitory concentration (IC50), concentration producing 50% inhibition of cell proliferation, of 1.3 μM (Keblys et al., 2004).

Ochratoxin A show an impact on the cytokine expression. An experiment on weaned pigs that ingested an OTA contaminated diet (181 ng/g of feed) has shown an increased level of TNF-alpha and IL-10 in plasma, with a decreased capacity to respond with cytokine expression to *ex vivo* challenge with lipopolysaccharides (LPS) ( Bernardini et al., 2014). By contrast, OTA has no effect on total and specific immunoglobulin concentrations (Harvey et al., 1992).

## 2.5. Zearalenone

Zearalenone is best known for its toxic effect on reproduction and fertility (Zinedine et al., 2007); it induces an estrogenic activity on animal (Fink-Gremmels and Malekinejad, 2007). Pigs are particularly sensitive to ZEN, which can induce edematous swelling and reddening of vulva, prolapse of the vulva, ovarian follicle damage and abortions (Schoevers et al., 2012 and Zinedine et al., 2007).

Only few papers described the effect of ZEN on immunity (Eriksen and Alexander, 1998). In pigs, exposure of intestinal epithelial cells ZEN (25 μM) has a tendency to increase the synthesis of the inflammatory cytokines IL-8 and IL-10 (Marin et al., 2015). Sows exposed to high concentration of ZEN (5–250 mg/kg feed or 200–1000 μg/kg BW per day) can develop a chronic inflammation of the genital tract (EFSA, 2011).

# 3. CONSEQUENCE OF MYCOTOXIN INDUCED IMMUNOMODULATION FOR PIG HEALTH

## 3.1. Susceptibility to infectious diseases

The broad immunosuppressive effect of mycotoxins may decrease host resistance to infectious diseases (Antonissen et al., 2014). Table 1 summarizes the data obtained in pigs.

In pigs, the consumption of feed contaminated with AF increases the severity of infection with *Erysipelothrix rhusiopathiae* (Cysewski et al., 1978). Similarly, during an experimental infection with *Brachyspira hydysenteriae*, the consumption of AF reduced the incubation time and increased the severity of diarrhea (Joens et al., 1981).

In presence of porcine circovirus type 2 (PCV2) virus, DON increases the severity of the viral infection, and in presence of the porcine reproductive and respiratory syndrome virus (PRRSV) it also increases the infection with more tissue lesions induced (Savard et al., 2014 and Savard et al., 2015b). During a bacterial infection, DON enhances the inflammatory reaction with an increased production of pro-inflammatory cytokines (Vandenbroucke et al., 2011). The elevation of circulating IgA in presence of low quantity of DON may, by contrast, increase the resistance to certain pathogens. Indeed, IgA initiates rapid and transient up-regulation of many immune related genes (Pestka et al., 2004).

In pigs, FB1 ingestion can induce intestinal infections, with some intestinal functions affected (Burel et al., 2013, Devriendt et al., 2009 and Oswald et al., 2003). The ingestion of FB1-contaminated feed was also associated with an increased susceptibility to pulmonary infection and an increase of the severity of the pathological changes with bacterial or viral pathogens (Devriendt et al., 2009, Halloy et al., 2005, Oswald et al., 2003, Posa et al., 2011, Posa et al., 2013 and Ramos et al., 2010).

**Table 1.** Influence of mycotoxins on susceptibility to infectious diseases in pig.

| Mycotoxin | Exposure dose | Exposure period | Pathogen | Effect compared with negative control | References |
|---|---|---|---|---|---|
| AFB1 | 0.07 and 0.14 mg/kg | 32 days | *Brachyspira hyodysentariae* | ↓ of incubation period for dysentery, ↑ diarrhea and dysentery time, ↑ death, visible clinical signs and lesions of dysentery at necropsy | Joens et al., 1981 |
| AF | 1.3 mg/kg feed | 25 days | *Ersypelothrix rhusiopathiae* | ↑ the severity of bacterial infection | Cysewski et al., 1978 |
| DON | 2.5 mg/kg feed | 3 weeks | PCV2 | ↑ viremia and lung viral load<br>no clinical effect | Savard et al., 2015b |
| DON | 3.5 mg/kg feed | 3 weeks | PRRSV | ↓ weight gain, ↑ lung lesions and mortality, no effect on viral replication | Savard et al., 2014 |
| DON | 1 µg/mL | 6 h | *Salmonella typhimurium* | synergistic ↑ gene expression *IL-12, TNF-α, IL-1β, IL-8, MCP-1 and IL-6* | Vandenbroucke et al., 2011 |
| T-2 toxin | 15 and 83 µg/kg feed | 23 days | *Salmonella typhimurium* | ↓ colonization of the cecum | Verbrugghe et al., 2012 |
| FB1 | 10 mg/kg feed | 3 days | *Bordetella bronchiseptica* & *Pasteurella multocida* (type D) | ↑ extent and severity of the pathological changes | Posa et al., 2011 |
| FB1 | 0.5 mg/kg BW | 6 days | *Escherichia coli* (SEPEC) | ↑ intestinal colonization; ↑ translocation to the mesenteric lymph node, lung, liver and spleen | Oswald et al., 2003 |
| FB1 | 1 mg/kg BW | 10 days | *Escherichia coli* (ETEC) | intestinal infection prolonged; impaired function of intestinal antigen presenting cells | Devriendt et al., 2009 |
| FB1 | 25.4 mg/kg feed | 42 days | *Mycoplasma hyopneumoniae* | ↑ severity of the pathological changes | Posa et al., 2013 |
| FB1 | 0.5 mg/kg BW | 7 days | *Pasteurella multocida* (type A) | ↓ growth rate and ↑ coughing; ↑ total number of cells, number of macrophages and lymphocytes in BALF; ↑ gross pathological lesions and histopathological lesion of lung | Halloy et al., 2005 |
| FB1 | 12 mg/kg BW | 18 days | PRRSV | ↑ histopathological lesions of lungs | Ramos et al., 2010 |
| FB1 | 11.8 mg/kg feed | 9 weeks | *Salmonella typhimurium* | Modification of the microbiota profiles | Burel et al., 2013 |
| OTA | 3 mg/kg feed | 3 weeks | *Brachyspira hyodysenteriae* & *Camphylobacter coli* | Salmonellosis arises spontaneously in animals fed the contaminated diet, clinical and patho-morphological changes (typical of salmonellosis), change of hematological and biological parameters | Stoev et al., 2000 |
| OTA | 75 µg/kg feed | 42 days | PCV2 | ↑ PCV2 replication in serum and tissues | Gan et al., 2015 |

AFB1 = aflatoxin B1; AF = aflatoxins; DON = deoxynivalenol; FB1 = fumonisin B1; OTA = ochratoxin A; BW = body weight; PCV2 = porcin circovirus type 2; PRRSV = porcine reproductive and respiratory syndrome virus.

Ingestion of OTA contaminated feed also increases susceptibility to natural infectious disease. Indeed, salmonellosis arose spontaneously in all piglets receiving an OTA contaminated diet, and when the animals were vaccinated against salmonellosis, the consumption of contaminated feed leads to spontaneous *Brachyspira hyodysenteriae*and *Campylobacter coli* infections ( Stoev et al., 2000). During a PCV2 infection, OTA increases the viremia in sera and tissues (Gan et al., 2015).

To the best of our knowledge, there are no data available concerning the effect of ZEN on*in vivo* analysis on mice which were fed 10 mg/kg ZEA (1.5 mg/kg BW per day) during 2 weeks, infected with *Listeria monocytogenes, and* showed a decreased resistance to*Listeria* with in increasing trend of the splenic bacterial counts, compared with control animals ( Pestka et al., 1987).

## 3.2. Reactivation of chronic infection

The effect of mycotoxin intoxication on the reactivation of chronic infection was also investigated. However, the experiment was not performed with pigs but with rodents. In the immunocompetent host, *Toxoplasma gondii* infection progresses to a chronic phase characterized by the presence of encysted parasites. Cyst rupture may occur, but infection remains latent and reactivation is prevented. In immunosuppressed animal and human subjects, such as patients infected with the human immunodeficiency virus, rupture is associated with the formation of new cysts and disease. Low and repeated doses of either AFB1 or T-2 toxin are able to accelerate *Toxoplasma* cyst rupture in previously infected mice ( Venturini et al., 1996).

## 3.3. Vaccination efficacy

Immunity acquired through vaccination can also be impaired by mycotoxin ingestion (Table 2). For example, AFB1 interferes with the development of acquired immunity in swine following erysipelas vaccination with bacterin preparation (a suspension of killed bacteria) of *E. rhusiopathiae* ( Cysewski et al., 1978). As already mentioned, ingestion of feed contaminated with AFB1 or T-2 Toxin reduced the vaccine response to the model antigen, ovalbumin, acting on the cellular and the humoral response respectively (Meissonnier et al., 2008a and Meissonnier et al., 2008b). Ingestion of low doses of another mycotoxin, FB1, decreases the specific antibody response mounted during*Mycoplasma* vaccination in pigs ( Taranu et al., 2005). In pigs exposed to OTA or FB1 and vaccinated against Aujesky disease (Suid Herpesvirus 1 [SuHV1]), the humoral immune response was greatly disturbed, with a strong decrease in antibody observed (Stoev et al., 2012). In diet contaminated with DON or FB1, pigs showed an alteration of the specific immune response upon vaccination with OVA (Grenier et al., 2011).

Likewise, feeding pigs a DON-contaminated diet was shown to inhibit the vaccination efficiency of PRRSV modified live vaccine by severely impairing viral replication (Savard et al., 2015a).

It should also been mentioned that the vaccine immune response is altered at mycotoxin doses that do not alter the global immune response (Meissonnier et al., 2008a,Meissonnier et al., 2008b and Taranu et al., 2005). The breakdown in vaccine immunity and may lead to the occurrence of disease even in properly vaccinated flocks. These reactions are of considerable consequence in animals for which we rely on an effective vaccination program for disease prevention.

# 4. THE PROBLEM OF MYCOTOXINS CO-CONTAMINATION

In the above paragraphs, the effects of single mycotoxin on immunity were described. However, mycotoxins often co-occur and animals are exposed to several mycotoxins at the same time. Indeed, raw materials can be contaminated by several fungi, which are able to simultaneously produce several mycotoxins, and in addition the diet of animal is composed of several commodities (Alassane-Kpembi et al., 2015 and Alassane-Kpembi et al., 2016). A worldwide survey on 7049 samples reported that 48% of feed and feedstuff samples are contaminated by 2 or more mycotoxins (Rodrigues and Naehrer, 2012). Other studies showed that 75%–100% of animal feed samples are contaminated with more than one mycotoxin (Guan et al., 2011a and Streit et al., 2012).

**Table 2.** Influence of mycotoxins on vaccination efficacy in pigs.

| Mycotoxin | Exposure dose | Antigen | Effect compared with negative control | References |
|---|---|---|---|---|
| AF | 1.3 mg/kg feed | *Erysipelothrix rhusiopathiae* | Interfered on the development of acquired immunity, | Cysewski et al., 1978 |
| AFB1 | 385–1807 µg/kg feed | OVA | Decreased and delayed cell-mediated immunity | Meissonnier et al., 2008b |
| DON | 3.5 mg DON/kg feed | OVA | Increased OVA-primary IgG antibody response | Lessard et al., 2015 |
| DON | 2.5–3.5 mg/kg BW | PRRSV | Decreased PRRSV post-vaccinal viremia and reduced vaccinal efficacy | Savard et al., 2015b |
| DON | 2.2–2.5 mg DON/kg feed | OVA | Increased concentration of OVA specific IgA and IgG | Pinton et al., 2008 |
| DON | 0.6–4.7 mg DON/kg | Human serum albumin, sheep red blood cells, paratuberculosis vaccine, tetanus toxoid and diphtheria toxoid | Significant dose-dependent reduction in secondary antibody response to tetanus toxoid | Overnes et al., 1997 |
| DON + ZEN | 2.1–3.2 mg DON/kg diet and 0.06–0.25 mg ZEN/kg diet | Parvovirus | No effect | Gutzwiller et al., 2007 |
| DON or FB1 | 3 mg DON/kg feed or 6 mg FB1/kg feed | OVA | Reduced anti-OVA antibody production with a decrease of lymphocytes proliferation | Grenier et al., 2011 |
| T-2 toxin | 1324–2102 µg/kg feed | OVA | Reduced anti-OVA antibody production without significant alteration to specific lymphocyte proliferation | Meissonnier et al., 2008a |
| FB1 | 8 mg/kg BW | *Mycoplasma agalactiae* | Decreased specific antibody titer | Taranu et al., 2005 |
| OTA | 1 mg/kg feed | *Salmonella choleraesuis* | Immunosuppression and delayed response to immunization | Stoev et al., 2000 |
| OTA or FB1 | 0.5 mg OTA/kg feed or 10 mg FB1/kg feed | Suid Herpesvirus 1 (Aujesky disease) | Decreased anti-SuHV1 antibody production after vaccination | Stoev et al., 2012 |

AF = aflatoxins; AFB1 = aflatoxin B1; DON = deoxynivalenol; ZEN = zearalenone; OTA = ochratoxin A; BW = body weight; OVA = ovalbumin; PRRSV = porcine reproductive and respiratory syndrome virus; OVA = ovalbumin.

The toxicity of mycotoxins mixtures cannot always be predicted based upon their individual toxicities. It can be antagonistic, additive or synergistic and increase the impact of each mycotoxin. The studies concerning the toxicity of mycotoxins mixture on pig immune response are scarce. Reduction of lymphocyte proliferation has been investigated in several pig *in vivo* studies, and different type of interaction were observed: additivity (co-exposure to AF and FB [0.05 and 30 mg/kg feed] for one month; co-exposure to OTA and T-2 toxin [2.5 and 8 mg/kg feed] for 30 days) or synergy (co-exposure to FB and DON [50 and 4 mg/kg feed] for 28 days) ( Grenier and Oswald, 2011). In animals co-exposed to DON and FB (6 and 3 mg/kg of feed) for 35 days, synergistic interaction was observed on lymphocytes proliferation upon mitogenic stimulation, additive interaction on cytokines expression (IL-8; IL-1β, IL-6 and macrophage inflammatory protein 1β) and antagonistic interaction on levels of specific IgA and cytokine expression (Grenier et al., 2011).

In animals co-exposed to DON and FB (6 and 3 mg/kg of feed) for 35 days, additive interaction on specific IgG, on lymphocytes proliferation upon mitogenic stimulation and on cytokines expression (IL-8, IL-1β, IL-6 and macrophage inflammatory protein 1β) was observed, and antagonistic interaction on levels of specific IgA was observed (Grenier et al., 2011).

# 5. CONCLUSION

Mycotoxins can contaminate many raw materials and cause significant health risk to animals. Numerous strategies are used to minimize mycotoxins contamination throughout the feed chain. In the fields, resistant crops associated as well as agronomic control measures can be used. Similarly, during feed storage and processing, physical, chemical and biological methods can reduce mycotoxin contamination. However once mycotoxins are present in feed, it's difficult to reduce their concentrations and their toxicity due to the stability of these compounds (Bryden, 2009). The simultaneous presence of several mycotoxins, not sensitive to the same detoxification procedure, also increases the difficulty to control animals' exposure to mycotoxins (Bryden, 2012). Recently, new detoxification biological methods showed that the use of bacteria (Grenier et al., 2012,Grenier et al., 2013 and Guan et al., 2011b), feed additives such as arginine or glutamate were effective to decrease the toxic effects of mycotoxins in young pigs (Duan et al., 2014,Wu et al., 2013 and Wu et al., 2015), even for exposition to mycotoxins mixtures (Yin et al., 2014 and Grenier et al., 2013).

Pig, a species very sensitive to mycotoxins, is really exposed due to a cereal rich diet. At the European level, regulation or recommendations exist for 6 mycotoxins that are often present in pig feed. They are FB, AF, OTA, DON, T-2/HT-2 toxins and ZEN. Exposure to these toxins induces several toxic effects on pig, including a modulation of the immune response. This later effect increases the susceptibility and severity of infectious diseases, and reduces the efficacy of vaccines. This is of particular note for animal husbandry because during infection, nutrients are used for the immune system instead of growth and development (Klasing, 2007). Consequently, mycotoxin contamination also has an indirect effect on animal productivity (Klasing, 2007 and Oswald et al., 2005).

The presence of new mycotoxins (emerging, masked, modified toxins, etc.) revealed by new analytical methods can also increase the risk for pig health. Currently, very few studies document the occurrence and toxicity of these toxins, thus there is a need to determine the risk they represent in pig production (Broekaert et al., 2015 and Pierron et al., 2016).

## CONFLICT OF INTEREST

The authors declare that they have no competing interests.

## ACKNOWLEDGMENT

Alix Pierron was supported by a fellowship from CIFRE (2012/0572, jointly financed by the BIOMIN Holding GmbH, Association Nationale de la Recherche Technique and INRA). This work was supported in part by the french project Tool4gutHealth, jointly financed by the Lallemand and the Région Midi-Pyrénées. Thanks are also due to Ryan Hines, Biomin, for language editing.

# REFERENCES

1. Alassane-Kpembi I, Puel O, Oswald IP. Toxicological interactions between the mycotoxins deoxynivalenol, nivalenol and their acetylated derivatives in intestinal epithelial cells. Arch Toxicol 2015;89:1337–46.

2. Alassane-Kpembi I, Schatzmayr G, Taranu I, Marin D, Puel O, Oswald IP. Mycotoxins co-contamination: methodological aspects and biological relevance of combined toxicity studies. Crit Rev Food Sci Nutr 2016.

3. Antonissen G, Martel A, Pasmans F, Ducatelle R, Verbrugghe E, Vandenbroucke V, et al. The impact of Fusarium mycotoxins on human and animal host susceptibility to infectious diseases. Toxins (Basel) 2014;6:430–52.

4. Bennett JW, Klich M. Mycotoxins. Clin Microbiol Rev 2003;16:497–516.

5. Bernardini C, Grilli E, Duvigneau JC, Zannoni A, Tugnoli B, Gentilini F, et al. Cellular stress marker alteration and inflammatory response in pigs fed with an ochratoxin contaminated diet. Res Vet Sci 2014;97:244–50.

6. Broekaert N, Devreese M, De Baere S, De Backer P, Croubels S. Modified Fusarium mycotoxins unmasked: from occurrence in cereals to animal and human excretion. Food Chem Toxicol 2015;80:17–31.

7. Bryden WL. Mycotoxins and mycotoxicoses: significance, occurrence and mitigation in the food chain. In: Ballantyne B, Marrs T, Syversen T, editors. General and applied toxicology. 3rd ed. Chichester, UK: John Wiley & Sons Ltd; 2009. p. 3529–53.

8. Bryden WL. Mycotoxin contamination of the feed supply chain: implications for animal productivity and feed security. Animal Feed Sci Technol 2012;173: 134–58.

9. Burel C, Tanguy M, Guerre P, Boilletot E, Cariolet R, Queguiner M, et al. Effect of low dose of fumonisins on pig health: immune status, intestinal microbiota and sensitivity to Salmonella. Toxins (Basel) 2013;5:841–64.

10. Chaytor AC, See MT, Hansen JA, de Souza AL, Middleton TF, Kim SW. Effects of chronic exposure of diets with reduced concentrations of aflatoxin and deoxynivalenol on growth and immune status of pigs. J Anim Sci 2011;89: 124–35.

11. Cysewski SJ, Wood RL, Pier AC, Baetz AL. Effects of aflatoxin on the development of acquired immunity to swine erysipelas. Am J Vet Res 1978;39:445–8.

12. Devriendt B, Gallois M, Verdonck F, Wache Y, Bimczok D, Oswald IP, et al. The food contaminant fumonisin B(1) reduces the maturation of porcine CD11R1(þ) intestinal antigen presenting cells and antigen-specific immune responses, leading to a prolonged intestinal ETEC infection. Vet Res 2009;40:40.

13. Drochner W, Schollenberger M, Piepho HP, Gotz S, Lauber U, Tafaj M, et al. Serum IgA-promoting effects induced by feed loads containing isolated deoxynivalenol (DON) in growing piglets. J Toxicol Environ Health A 2004;67:1051–67.

14. Duan J, Yin J, Wu M, Liao P, Deng D, Liu G, et al. Dietary glutamate supplementation ameliorates mycotoxin-induced abnormalities in the intestinal structure and expression of amino acid transporters in young pigs. PLoS One 2014;9: –112357.

15. EFSA. Scientific Opinion on the risks for public health related to the presence of zearalenone in food. EFSA J 2011;9:2197.

16. Eriksen GS, Alexander JA. In: Nordic Council of Miniesters, editor. Fusarium Toxins in Cereal e A Risk assessment, vol. 502. Copenhagen: Tema Nord; 1998. p. 7–58.

17. Fink-Gremmels J, Malekinejad H. Clinical effects and biochemical mechanisms associated with exposure to the mycoestrogen zearalenone. Anim Feed Sci Technol 2007;137:326–41.

18. Gan F, Zhang Z, Hu Z, Hesketh J, Xue H, Chen X, et al. Ochratoxin A promotes porcine circovirus type 2 replication in vitro and in vivo. Free Radic Biol Med 2015;80: 33–47.

19. Grenier B, Loureiro-Bracarense AP, Lucioli J, Pacheco GD, Cossalter AM, Moll WD, et al. Individual and combined effects of subclinical doses of deoxynivalenol and fumonisins in piglets. Mol Nutr Food Res 2011;55:761–71.

20. Grenier B, Oswald IP. Mycotoxin co-contamination of foods and feeds: metaanalysis of publications describing toxicological interactions. World Mycotoxin J 2011;4:285–313.

21. Grenier B, Bracarense AP, Schwartz HE, Trumel C, Cossalter AM, Schatzmayr G, et al. The low intestinal and hepatic toxicity of hydrolyzed fumonisin B(1) correlates with its inability to alter the metabolism of sphingolipids. Biochem Pharmacol 2012;83:1465–73.

22. Grenier B, Bracarense AP, Schwartz HE, Lucioli J, Cossalter AM, Moll WD, et al. Biotransformation approaches to alleviate the effects induced by fusarium mycotoxins in swine. J Agric Food Chem 2013;61:6711–9.

23. Guan S, Gong M, Yin Y, Huang R, Ruan Z, Zhou T, et al. Occurrence of mycotoxins in feeds and feed ingredients in China. J Food, Agric Environ 2011a;9:163–7.

24. Guan S, Zhou T, Yin Y, Xie M, Ruan Z, Young JC. Microbial strategies to control aflatoxins in food and feed. World Mycotoxin J 2011b;4:413–24.

25. Gutzwiller A, Czegledi L, Stoll P, Bruckner L. Effects of Fusarium toxins on growth, humoral immune response and internal organs in weaner pigs, and the efficacy of apple pomace as an antidote. J Anim Physiol Anim Nutr 2007;91:432–8.

26. Halloy DJ, Gustin PG, Bouhet S, Oswald IP. Oral exposure to culture material extract containing fumonisins predisposes swine to the development of pneumonitis caused by Pasteurellamultocida. Toxicology 2005;213:34–44.

27. Harvey RB, Elissalde MH, Kubena LF, Weaver EA, Corrier DE, Clement BA. Immunotoxicity of ochratoxin A to growing gilts. Am J Vet Res 1992;53:1966–70.

28. Joens LA, Pier AC, Cutlip RC. Effects of aflatoxin consumption on the clinical course of swine dysentery. Am J Vet Res 1981;42:1170–2.

29. Keblys M, Bernhoft A, Hofer CC, Morrison E, Larsen HJ, Flaoyen A. The effects of the Penicillium mycotoxins citrinin, cyclopiazonic acid, ochratoxin A, patulin, penicillic acid, and roquefortine C on in

vitro proliferation of porcine lymphocytes. Mycopathologia 2004;158:317–24.

30. Klasing KC. Nutrition and the immune system. Br Poult Sci 2007;48:525–37.

31. Lessard M, Savard C, Deschene K, Lauzon K, Pinilla VA, Gagnon CA, et al. Impact of deoxynivalenol (DON) contaminated feed on intestinal integrity and immune response in swine. Food Chem Toxicol 2015;80:7–16.

32. Liu BH, Yu FY, Chan MH, Yang YL. The effects of mycotoxins, fumonisin B1 and aflatoxin B1, on primary swine alveolar macrophages. Toxicol Appl Pharmacol 2002;180:197–204.

33. Maresca M, Fantini J. Some food-associated mycotoxins as potential risk factors in humans predisposed to chronic intestinal inflammatory diseases. Toxicon 2010;56:282–94.

34. Marin DE, Taranu I, Bunaciu RP, Pascale F, Tudor DS, Avram N, et al. Changes in performance, blood parameters, humoral and cellular immune responses in weanling piglets exposed to low doses of aflatoxin. J Anim Sci 2002;80:1250–7.

35. Marin DE, Taranu I, Pascale F, Lionide A, Burlacu R, Bailly JD, et al. Sex-related differences in the immune response of weanling piglets exposed to low doses of fumonisin extract. Br J Nutr 2006;95:1185–92.

36. Marin DE, Motiu M, Taranu I. Food contaminant zearalenone and its metabolites affect cytokine synthesis and intestinal epithelial integrity of porcine cells. Toxins (Basel) 2015;7:1979–88.

37. Mehrzad J, Devriendt B, Baert K, Cox E. Aflatoxin B(1) interferes with the antigenpresenting capacity of porcine dendritic cells. Toxicol Vitro 2014;28:531–7.

38. Mehrzad J, Devriendt B, Baert K, Cox E. Aflatoxins of type B and G affect porcine dendritic cell maturation in vitro. J Immunotoxicol 2015;12:174–80.

39. Meissonnier GM, Marin DE, Galtier P, Bertin G, Taranu I, Oswald IP. Modulation of the immune response by a group of fungal food contaminant, the aflatoxins. In: Mengheri E, Roselli M, Bretti MS, Finamore A, editors. Nutrition and immunity; 2006. p. 147–66.

40. Meissonnier GM, Laffitte J, Raymond I, Benoit E, Cossalter AM, Pinton P, et al. Subclinical doses of T-2 toxin impair acquired immune response and liver cytochrome P450 in pigs. Toxicology 2008a;247:46–54.

41. Meissonnier GM, Pinton P, Laffitte J, Cossalter AM, Gong YY, Wild CP, et al. Immunotoxicity of aflatoxin B1: impairment of the cell-mediated response to vaccine antigen and modulation of cytokine expression. Toxicol Appl Pharmacol 2008b;231:142–9.

42. Oswald IP, Desautels C, Laffitte J, Fournout S, Peres SY, Odin M, et al. Mycotoxin fumonisin B1 increases intestinal colonization by pathogenic Escherichia coli in pigs. Appl Environ Microbiol 2003;69:5870–4.

43. Oswald IP, Marin DE, Bouhet S, Pinton P, Taranu I, Accensi F. Immunotoxicological risk of mycotoxins for domestic animals. Food Addit Contam 2005;22:354–60.

44. Overnes G, Matre T, Sivertsen T, Larsen HJ, Langseth W, Reitan LJ, et al. Effects of diets with graded levels of naturally deoxynivalenol-contaminated oats on immune response in growing pigs. Zentralbl Veterinarmed A 1997;44:539–50.

45. Pestka JJ, Tai JH, Witt MF, Dixon DE, Forsell JH. Suppression of immune response in the B6C3F1 mouse after dietary exposure to the Fusarium mycotoxins deoxynivalenol (vomitoxin) and zearalenone. Food Chem Toxicol 1987;25:297–304.

46. Pestka JJ, Zhou HR, Moon Y, Chung YJ. Cellular and molecular mechanisms for immune modulation by deoxynivalenol and other trichothecenes: unraveling a paradox. Toxicol Lett 2004;153:61–73.

47. Pestka JJ. Deoxynivalenol: mechanisms of action, human exposure, and toxicological relevance. Arch Toxicol 2010;84:663–79.

48. Pierron A, Mimoun S, Murate LS, Loiseau N, Lippi Y, Bracarense AF, et al. Intestinal toxicity of the masked mycotoxin deoxynivalenol-3-beta-D-glucoside. Arch Toxicol 2016.

49. Pinton P, Accensi F, Beauchamp E, Cossalter A-M, Callu P, Grosjean Fo, et al. Ingestion of deoxynivalenol (DON) contaminated feed alters the pig vaccinal immune responses. Toxicol Lett 2008;177:215–22.

50. Posa R, Donko T, Bogner P, Kovacs M, Repa I, Magyar T. Interaction of Bordetella bronchiseptica, Pasteurella multocida, and fumonisin B1 in the porcine respiratory tract as studied by computed tomography. Can J Vet Res 2011;75:176–82.

51. Posa R, Magyar T, Stoev SD, Glavits R, Donko T, Repa I, et al. Use of computed tomography and histopathologic review for lung lesions produced by the interaction between Mycoplasma hyopneumoniae and fumonisin mycotoxins in pigs. Vet Pathol 2013;50:971–9.

52. Rafai P, Tuboly S, Bata A, Tilly P, Vanyi A, Papp Z, et al. Effect of various levels of T-2 toxin in the immune system of growing pigs. Vet Rec 1995;136:511–4.

53. Ramos CM, Martinez EM, Carraso AC, Puente JHL, Quezada F, Perez JT, et al. Experimental trial of the effect of fumonisin B1 and PRRS virus in swine. Vet Adv 2010;9:1301–10.

54. Rodrigues I, Naehrer K. A three-year survey on the worldwide occurrence of mycotoxins in feedstuffs and feed. Toxins (Basel) 2012;4:663–75.

55. Savard C, Pinilla V, Provost C, Gagnon CA, Chorfi Y. In vivo effect of deoxynivalenol (DON) naturally contaminated feed on porcine reproductive and respiratory syndrome virus (PRRSV) infection. Vet Microbiol 2014;174:419–26.

56. Savard C, Gagnon CA, Chorfi Y. Deoxynivalenol (DON) naturally contaminated feed impairs the immune response induced by porcine reproductive and respiratory syndrome virus (PRRSV) live attenuated vaccine. Vaccine 2015a;33:3881–6.

57. Savard C, Provost C, Alvarez F, Pinilla V, Music N, Jacques M, et al. Effect of deoxynivalenol (DON) mycotoxin on in vivo and in vitro porcine circovirus type 2 infections. Vet Microbiol 2015b;176:257–67.

58. Schoevers EJ, Santos RR, Colenbrander B, Fink-Gremmels J, Roelen BA. Transgenerational toxicity of Zearalenone in pigs. Reprod Toxicol 2012;34:110–9.

59. Silvotti L, Petterino C, Bonomi A, Cabassi E. Immunotoxicological effects on piglets of feeding sows diets containing aflatoxins. Vet Rec 1997;141:469–72.

60. Stockmann-Juvala H, Savolainen K. A review of the toxic effects and mechanisms of action of fumonisin B1. Hum Exp Toxicol 2008;27:799–809.

61. Stoev SD, Goundasheva D, Mirtcheva T, Mantle PG. Susceptibility to secondary bacterial infections in growing pigs as an early response in ochratoxicosis. Exp Toxicol Pathol 2000;52:287–96.

62. Stoev SD, Gundasheva D, Zarkov I, Mircheva T, Zapryanova D, Denev S, et al. Experimental mycotoxic nephropathy in pigs provoked by a mouldy diet containing ochratoxin A and fumonisin B1. Exp Toxicol Pathol 2012;64:733–41.

63. Streit E, Schatzmayr G, Tassis P, Tzika E, Marin D, Taranu I, et al. Current situation of mycotoxin contamination and co-occurrence in animal feedefocus on Europe. Toxins (Basel) 2012;4:788–809.

64. Streit E, Naehrer K, Rodrigues I, Schatzmayr G. Mycotoxin occurrence in feed and feed raw materials worldwide: long-term analysis with special focus on Europe and Asia. J Sci Food Agric 2013;93:2892–9.

65. Swamy HV, Smith TK, MacDonald EJ, Karrow NA, Woodward B, Boermans HJ. Effects of feeding a blend of grains naturally contaminated with Fusarium mycotoxins on growth and immunological measurements of starter pigs, and the efficacy of a polymeric glucomannan mycotoxin adsorbent. J Anim Sci 2003;81:2792–803.

66. Taranu I, Marin DE, Bouhet S, Pascale F, Bailly JD, Miller JD, et al. Mycotoxin fumonisin B1 alters the cytokine profile and decreases the vaccinal antibody titer in pigs. Toxicol Sci 2005;84:301–7.

67. Vandenbroucke V, Croubels S, Martel A, Verbrugghe E, Goossens J, Van Deun K, et al. The mycotoxin deoxynivalenol potentiates intestinal inflammation by Salmonella typhimurium in porcine ileal loops. PLoS One 2011;6: –23871.

68. Venturini MC, Quiroga MA, Risso MA, Lorenzo CD, Omata Y, Venturini L, et al. Mycotoxin T-2 and aflatoxin B1 as immunosuppressors in mice chronically infected with Toxoplasma gondii. J Comp Pathol 1996;115:229–37.

69. Verbrugghe E, Vandenbroucke V, Dhaenens M, Shearer N, Goossens J, De Saeger S, et al. T-2 toxin induced Salmonella typhimurium intoxication results in decreased Salmonella numbers in the cecum contents of pigs, despite marked effects on Salmonella-host cell interactions. Vet Res. 2012;43:22.

70. Weaver AC, See MT, Hansen JA, Kim YB, De Souza AL, Middleton TF, et al. The use of feed additives to reduce the effects of aflatoxin and deoxynivalenol on pig growth, organ health and immune status during chronic exposure. Toxins (Basel) 2013;5:1261–81.

71. Wild CP, Gong YY. Mycotoxins and human disease: a largely ignored global health issue. Carcinogenesis 2010;31:71–82.

72. Wu L, Wang W, Yao K, Zhou T, Yin J, Li T, et al. Effects of dietary arginine and glutamine on alleviating the impairment induced by deoxynivalenol stress and immune relevant cytokines in growing pigs. PLoS One 2013;8: –69502.

73. Wu L, Liao P, He L, Feng Z, Ren W, Yin J, et al. Dietary L-arginine supplementation protects weanling pigs from deoxynivalenol-induced toxicity. Toxins (Basel) 2015;7:1341–54.

74. Yin J, Ren W, Duan J, Wu L, Chen S, Li T, et al. Dietary arginine supplementation enhances intestinal expression of SLC7A7 and SLC7A1 and ameliorates growth depression in mycotoxin-challenged pigs. Amino Acids 2014;46:883–92.

75. Zinedine A, Soriano JM, Molto JC, Manes J. Review on the toxicity, occurrence, metabolism, detoxification, regulations and intake of zearalenone: an oestrogenic mycotoxin. Food Chem Toxicol 2007;45:1–18.

# CHAPTER 6

# Animal Models for Study of Diabetes Mellitus

**A. Lukačínová[1], B. Hubková[2], O. Rácz[3, 4] and F. Ništiar[3]**

[1] Department of Physiology, Faculty of Medicine, Šafárik University, Košice, Slovak Republic

[2] Department of Medical and Clinical Biochemistry, Faculty of Medicine, Šafárik University, Košice, Slovak Republic

[3] Department of Pathological Physiology, Faculty of Medicine, Šafárik University, Košice, Slovak Republic

[4] Department of Nanobiotechnology and Regenerative Medicine, Faculty of Healthcare, Miskolc University, Hungary

## 1. INTRODUCTION

Diabetes mellitus is a heterogeneous group of chronic disorders of carbohydrate, lipid and protein metabolism characterized by high blood glucose levels due to relative or absolute deficiency of insulin (Eiselein et al., 2004). Hyperglycemia, the primary clinical manifestation of diabetes, is associated with the development of diabetic complications. Several studies have suggested that hyperglycemia accelerates the development of chronic complications via several mechanisms, including increased aldose reductase related polyol pathway flux, increased formation of advanced glycation end-products (AGEs), activation of protein kinase C isoforms, increased hexosamine pathway flux, and overproduction of reactive forms of oxygen (Brownlee, 2001). AGEs are a group of complex and heterogeneous compounds, including brown and fluorescent cross-linking substances (e.g., pentosidine), non-fluorescent cross-linking products (e.g., methylglyoxal lysine dimers), or non-fluorescent, non-cross linking adducts (e.g., carboxymethyl lysine) (Dyer et al., 1991). Increasing evidence identifies AGE formation as the critical pathogenic link between hyperglycemia and long-term complications of diabetes: nephropathy, neuropathy, and retinopathy (Wada & Yagihashi, 2005). Therefore, another mode of diabetes treatment independent of blood glucose levels, inhibition of AGE formation, could be useful in the prevention or reduction of certain diabetic complications (Dong et al., 2010) in both main forms of the illness, Type 1 diabetes mellitus (T1D, insulin-dependent diabetes mellitus, IDDM) and Type 2 diabetes mellitus (T2D, noninsulin-dependent diabetes mellitus, NIDDM), and also in secondary forms related to gestation or other disorders.

In 2010, WHO estimated that 285 million people were living with diabetes (corresponding to 6.4% of the world's adult population). About 7 million people develop the disease each year and 3.9 million deaths were attributed to diabetes yearly (Shaw et al., 2010; Roglic & Unwin, 2010). Current predictions estimate that the prevalence of diabetes will reach 438 million by 2030 (corresponding to 7.8% of the adult population) and that 80% of prevalent cases will occur in the developing world (Roglic & Unwin, 2010). The increase is mainly driven by changes in dietary habits and low levels of physical activity (Wild et al., 2004). In the poorest countries, diabetes is more common among the better-off, but economic development quickly reverses this trend so that people from lower socioeconomic groups are more affected by diabetes; consequences are worse among the poor in all countries (Whiting et al., 2010, Blas & Sivasankara Kurup, 2010). Diabetes mellitus and specially diabetes with chronic complications belongs to the diseases requiring huge costs (Ettaro et al., 2004).

Diabetes mellitus is a syndrome characterized by many symptoms most typical of which is hyperglycemia. T1D is 10% of all diabetes cases, but its prevalence is constantly increasing (Green et al., 2001). This type occurs most frequently around the age of 14 years and individuals affected by this type must be treated throughout life with insulin injections (Rossini et al., 2003). Disease is the result of inflammatory islet infiltration (insulitis) and selective destruction of insulin producing - cells (Atkinson and Eisenbarth, 2001). In general, diabetes being an autoimmune basis, T1D occurs frequently in individuals with other autoimmune diseases, especially intestinal or thyroid gland. It is strongly bound to major histocompatibility system (MHC), is dependent on T cells and can be modified by immunosuppression. When you start an autoimmune mechanism, exogenous factors play a role particularly of viral origin, but also antigens of the host cells based on molecular mimicry. These trigger tissue-specific immune response producing cross-reactive effector cells or antibodies that recognize self-proteins of cells of the pancreas (Kukreja & Maclaren, 1999). Coxsackie B4 virus, for example, contains a sequence of 18 amino acids similar to enzyme glutamic acid decarboxylase (GAD) of pancreatic -cells (Kaufman et al., 1992). At the same time viruses can cause subtle damage to the -cells followed by autoimmune response against damaged -damaged cells with sequestrated virus antigen. Other studies have identified endogenous retroviral genomes in diabetic islets, but whether the virus initiates or is only a marker of the disease remains unclear (Benoist & Mathis, 1997). Regardless of the type of trigger are activated specific self reactive T cell clones against pancreatic -cells, which then infiltrate the islets of the pancreas. It is believed that these T cell clones belonging to the T helper 1 (Th1) subset. These Th1 cells produce characteristic cytokines, such as IFN and IL-2, which are considered as triggers of insulitis and destruction of -cells of the pancreas (Suarez-Pinzon & Rabinovitch, 2001). Following the massive loss of pancreatic -cells develop severe deficiency of insulin resulting in hyperglycemia. This causes glycogenolysis in the liver, activate gluconeogenetic pathways and decreases cellular uptake of glucose in peripheral tissues (muscle and adipose). Extensive degradation of fats and oxidation of fatty acids leads to hyperlipidemia and ketosis. The essential symptoms include hyperglycemia, polyuria due to osmotic diuresis, thirst due to hyperosmolar state, weight loss due to depletion of fat reserves and negative nitrogen balance, neurotoxicity due to hyperglycemia and ketoacidosis (Eiselein et al., 2004). If patients are not treated die due to circulatory collapse and coma. When initiating therapy by exogenous insulin, but are frequent hyperglycemia, develop micro- and macroangiopathies. Therefore, in diabetic patients is increased risk for coronary heart disease, stroke, gangrene of the lower limbs, chronic kidney disease, blindness and visual disturbances, and autonomic and peripheral neuropathy. These problems reduce life expectancy by up to 25% (Williams & Pickup, 2004) and the most common causes of death are diseases of heart and kidney.

In addition to people diabetes mellitus are quite often found in animals living with man (dog, cat), pets (or livestock) and laboratory animals (Malaisse & Sener, 2008).

For eliminate ethical and logistic problems in the study of T1D in connection with the heterogeneity of outbreed human populations, have been developed so many animal models of induced and spontaneous diabetes with the possibility of more frequent sampling, biopsy and autopsy samples. In these models it is possible to breeding at a controlled heredity and study of various environmental factors in relatively large and uniform populations. Animal models have a special status in the study of pathogenesis of chronic complications of diabetes, which are considered the following mechanisms: (1) non-enzymatic glycosylation, (2) intracellular hyperglycemia with associated disturbances in the polyol pathway, (3) activation of protein kinase C (PKC) and (4) increased hexosamine pathway flux (Eiselein et al., 2004). Non-enzymatic glycosylation lead to irreversible formation of advanced glycation end products (AGEs) through the Maillard reaction, and this is probably the best studied pathogenetic factor of diabetic complications. Pathological effects of AGEs are induced by impaired function of glycosylated products and by activation of AGE receptors on endothelial cells, monocytes, macrophages, lymphocytes and mesenchymal cells. Increased risk of heart diseases in diabetics is mainly attributed to the formation of AGEs (Eiselein et al., 2004). Glycosylation of collagen type IV in basement membranes of blood vessels leads to cross-links between interstitial proteins and lipoproteins, e.g., LDL. (Vlassara, 1996). LDL can be glycosylated and subsequently oxidized (Bucala et al., 1993). Modified LDL can then be bound to receptors of macrophages (CD36). The resulting foam cell formation and development of the fatty streak in the sub-endothelial space is the beginning of the atherosclerotic process (Ohgami et al., 2001). Since in diabetics the levels of AGEs of apoprotein B and phospholipids are several times higher as compared to nondiabetics, diabetics have a three- to four-fold higher risk of cardiovascular diseases (Bucala et al., 1994) and it is expected that non-enzymatic glycation is responsible also for vascular occlusions (Peppa et al., 2004). Other studies have attributed the importance of AGEs in hypertension, kidney pathology and erectile dysfunction in diabetics. These injuries are caused by AGEs present in the vascular matrix, where it inhibits the vasodilatory effect of EDRF (NO) and increased expression of endothelin-1, a potent vasoconstrictor (Quehenberger et al., 2000). Many of the effects of AGEs are dependent on the receptor. The best characterized receptor of AGEs is the receptor for advanced glycation end products (RAGE), which is a member of the immunoglobulin superfamily of cell membrane molecules (Stern et al., 2002). Studies in rodent models demonstrated that blocking RAGE can inhibit vascular hyperpermeability and reduced development of atherosclerotic lesion (Wendt et al., 2002). These results suggest that AGE-RAGE system may be a promising target for prophylaxis and treatment of late complications of diabetes mellitus. Hyperglycemia can cause various damages by changing polyol pathway, diacylglycerol-PKC pathway is important for develop of micro- and macro-angiopathy, and finally hexosamine biosynthetic pathway is also involved in the development of diabetic vascular complications (Eiselein et al., 2004).

In the study of T1D is primarily used two species of laboratory animals, rat and mouse (von Herrath and Nepom, 2009). The most commonly used model is non-obese diabetic (NOD) mouse (Serreze and Leiter, 2001). Many data in the literature explains the genetics and immunology of these animals and they were identified by at least 27 genetic loci and many immunological defects (Serreze and Leiter, 2001). Their disadvantages as a model consisted of two facts. Firstly, for this model in addition to immunosuppression also other immunomodulatory interventions showed a preventive effect on diabetes (Atkinson & Leiter, 1999), in contrast, these were ineffective in humans (Skyler et al., 2002). Second, because some of the negative factors of environment, particularly viral infections (Leiter, 1998) reduced the frequency of diabetes and often act preventively in mice (Atkinson & Leiter, 1999), while in humans viruses are known triggers. Although the original NOD mice and congenic strains derived from them provide valuable information and is necessary to keep in mind that they are not completely appropriate alternative to studying human T1D (Greiner et al., 2001). Rats provide another model. Such are BB rats (Mordes et al., 2001), a widely studied model susceptible to autoimmune destruction of pancreatic -cells. In some experiments also nonmammal models can be exploited due they easy handling and low economic costs.

Use of animal models in diabetes research has a long tradition. Our aim was to present an overview of animal models used in research of diabetes mellitus and highlight some technological and scientific problems associated with interpretation of carrying out these experiments in accordance with current European legislation on animal protection and exploitation of animals for experimental purposes (Ništiar et al., 2006). It is based on detailed analysis of the literature, using well-known databases such as MEDLINE, HighWire Press, PubMed and basic book, "Animal Models of Diabetes: A Primer" (Sima & Shafrir, 2000).

Already in 1890, von Mehring and Minkowski induced in dogs by removing the pancreas acute diabetes mellitus (Minkowski, 1989). In addition to partial and the total pancreatectomy, there are also non-surgical methods to induce hyperglycemia. There are five groups of diabetogenic substances: chemical and biological agents, potentiators, peptides and steroids (van der Werf et al., 2007). Among the animal models include in particular rodents, which are especially suitable for their low cost, short generation time, inherited forms of obesity and hyperglycemia. There are known animal models to study T1D also T2D

(Rees & Alcolado, 2005; Yang & Santamaria, 2006). Animals with spontaneous T1D have been drawn by inbreed breeds in different laboratories. In contrast, animal models of T2D are very heterogeneous. This includes not only animals with single-gene mutation, but also other types with insulin resistant syndrome and impaired pancreatic -cells (Matteucci & Giampietro, 2008).

More recently techniques using methods of molecular biology have produced genetically modified mouse models, including knockout and transgenic animals. Knockout animals have been defective gene in embryonic stem cells. In transgenic animals modified gene is incorporated into the pronucleus of zygote and then randomly into the genome of the animal and is transferred to the offspring.

Diabetic animal models undoubtedly have enormous benefit in clarifying the effect of insulin and insulin therapy. They also have some shortcomings, particularly when extrapolating results to humans and diabetic complications in humans (Tesch & Nikolic-Paterson, 2006) as well as some results may be misleading in the study of T1D prevention by using of model rodents (Leiter & von Herrath, 2004). It is claimed that the rodents in this area do not adequately reflect the situation in humans (Yang & Santamaria, 2006). Accordingly, these models and experiments will be necessary to standardize especially for prevention studies. Reliable results about the differences of pharmaceutical efficiency or survival of animals, delays and onset of disease, temporal relation of events require a very clear interpretation and repeatability of the experiment as well as the establishment of appropriate databases.

## 2. BASIC CONSIDERATIONS

Already in the Helsinki Declaration of the World Medical Association, 1964 (since been revised and amended) Rule 12 says that "to be secured by appropriate conditions, environment and care of experimental animals included in the experiment." The protection and welfare (only vertebrates) used for experimental purposes is regulated by: *Council Directive 86/609/EEC*, *Directive 2003/65/EC*,*Council Decision 1999/575/EC* and *Council Decision 2003/584/EC*. While there are considerable differences in the application of these measures in own legislation of individual countries. 23$^{rd}$ January 2006 the European Commission presented a modified and a new *Action Plan* for to ensure the welfare of animals in the future to a much higher level in all European Union member countries.

In 1959, zoologist William Russell and microbiologist Rex Burch presented a proposal for research on the three R: replacement, reduction and refinement (Russell & Burch, 1959). As the replacement is considered using any technique with insensitive material that replaces the use of susceptible live vertebrate animals. For the relative replacement is considered when animals are used only in certain parts of experiment thereby minimize distress during the whole experiment. Absolute replacement is if susceptible live animals are not used at any stage of the experiment. Organ and tissue cultures represent a transition between absolute and relative replacement. Reduction is one of the methods that allow researchers to reduce the number of animals used for research without reducing the statistical significance. In terms of statistical significance is extremely important to determine the number of animals needed for certain types of biomedical research (Dell et al., 2002). The significance of the test must be such as to enable to assess the clinical significance of the monitored phenomenon. Although the difference is less, we can be made certain conclusions (trends) without the need to increase the number of animals. This is necessary to meet two basic requirements, and that the differences have a normal distribution to determine the standard deviation.

Refinement is provided by better care of the animals. Fourth R, responsibility aims at increasing the accountability of scientists using animals in experiments (Bark, 1995). This area is in the forefront in recent years.

*Report A2005* submitted consensus on animal experiments in the present. It also submitted a proposal for the basic procedures and compliance with conditions of four R (Perry, 2007).

Regarding diabetes mellitus is necessary to admit that the conditions regarding the extrapolation of the results of animal experiments to conditions in humans are not clearly clarified. There are neither clearly intended technical conditions for carrying animal experiments for obtaining scientific knowledge. If the technological of the performance of experiments conditions are not sufficiently clarified one cannot even assume that their moral and ethical aspects are in order. From that aspect, it is clear only that the animals were subjected to experiments at the lowest discomfort (Rees & Alcolado, 2005).

Animal models are possible principally to divide into five groups (Hau, 2008):

1) induced (experimental) models;

2) spontaneous (genetic mutant) models;

3) genetically modified models (transgenic animals);

4) negative models (strain of animals in whom the studied disease does not occur) and

5) orphan models, describing the malfunctions that occur in the model animal, but do not occur in humans (e.g., Marek's disease, bovine spongiform encephalopathy, etc.).

Induced (experimental) models are represented by animals in which a modeled phenomenon as diabetes was induced by certain agents, e.g., chemical (alloxan, streptozotocin, cow's milk), viruses (encephalo-myocarditis virus) or pancreatectomy.

The use of animals for experimental purposes is always encountered negative response from the public, especially on the basis of ethical and religious reasons. These efforts led to the gradual formation of legislation towards the protection of animals used for experimental purposes. Nevertheless these experiments have contributed significantly to the current scientific knowledge of human biology, physiology, endocrinology and pharmacology (Loew, 1996). Outputs from these experiments were often not extrapolated to humans and it has also led to efforts to reduce them and to legislation guidance and control. On the other hand it should be noted that these very often contribute to a better understanding of many biological phenomena. Science on laboratory animals can be defined as a professional field focused on issues of scientific, ethical and lawful use of animals for biomedical research, i.e. interdisciplinary science involving biological and pathobiological details for optimal scientific use of animals as models for humans or other species. In general it deals with the quality of animals as susceptible objects in biomedical research. It includes comparative biology of laboratory animals, aspects of the breeding and reproduction (cross), welfare, economy of farming, anesthesia, euthanasia[1] - , and experimental procedures. The basic precondition for the use of animals for experimental purposes is a competence of researcher, including a solid knowledge about the biological needs, care and handling of animals.

If the use of animals for experimental purposes is an essential condition a similarity of the animal with modeled objects in the light of studied phenomenon, i.e. concept the analogy of animal models. If the analogy is closer the possibility of extrapolation of results is much more reliable. Extrapolation is used to express to what extent can be used the results from animal experiments to humans (how are applicable to humans). In the analysis from big multinational pharmaceutical companies in 150 compounds were monitored compliance (concordance) between animal models and human subjects (Olson et al., 2000). Concordance determining the toxicity of the substances to humans if were tested in rodents and non-rodents was 71%, when used only non-rodents 63% and if used only rodents 43%. High concordances are detected in the cardiovascular toxicity (80%), hematological toxicity (91%) and gastrointestinal toxicity (85%). Low concordances are in the neurological manifestations. Despite the high concordances in animal experiments are often reports of damages in humans by preparations of certain pharmaceutical companies. E.g., penicillin is fatal for the guinea pig, but it is well tolerated by humans, aspirin is teratogenic for cats, dogs, guinea pigs, rats, mice and monkeys, but not teratogenic for pregnant women, despite their frequent use (Mann, 1984). Thalidomide causing malformations in 10 000 children did not cause birth defects in rats (Koppányi & Avery, 1966) nor in many other species (Miller & Strömland, 1999). The close phylogenetic or morphological similarity is not crucial for the biochemical mechanisms and physiological responses, although in many cases this is so (Beynen & Hau, 2001). A very important difference between experimental animals and human populations is their genetic variability. Experimental animals are genetically almost identical in contrast to man exhibiting great variability. It is therefore possible to lay down precise rules for extrapolation of results from one species to another species, although in the literature have been made to certain procedures (Calabrese, 1991). Under the extrapolation is necessary to bear in mind certain mathematically expressed values, although it often looks precisely from this aspect, e.g., in determining the toxicity of certain substances or to determine of pharmacologically effective doses of drugs. In our opinion, fundamental importance is the rather in detection of toxic or therapeutic efficacy of substances. For specific determination of toxic or biologically effective dose, they should be taken in extrapolating only as a possible benchmark. Therefore, for any type of research should be borne in mind that the use of animal model does not attempt to extrapolate the main objective of the immediate outcome of the man, but looking for an answer to questions of researcher. While examining the need of experiments on animals it is first necessary to determine whether the experiment gave relevant answer to the experimenter, and whether the answer to the question enriches the current knowledge about the studied phenomenon. With this reason should be hypothesis (question) subjected to analysis before the experiment (approval of animal experiments, ethics committees, input opponency), and after publishing the results (answers to questions) and by the public opinion (responses).

Selection of animal species for the experimental purpose must be based on the greatest similarity between model object and modeling object, therefore the found results showed the strong concordance and therefore extrapolation was valid. It is well known that the metabolic rate in young and small animals is much higher than in large and old animals, because body size is the one of the key indicators and benchmarks for extrapolation, and precisely on the basis that have been submitted to the methods of extrapolation for the calculation of effective or toxic doses of various compounds (Hau & Poulsen, 1988).

# 3. ANIMAL MODELS FOR THE STUDY OF DIABETES MELLITUS

Recently almost 140 years passed from the excellent experiments of Mehring and Minkowski with pancreatectomized dogs. There have been many experiments with this model on rabbits and dogs when Banting and Best came to experiment, which led to the beginning of the isolation and purification of insulin (Bliss, 1982). Dog Marjorie, who was first treated with exogenous insulin, is probably one of the most famous experimental animal in the history, comparable only with Dolly the sheep from genetic studies at the end of the last century. The early experimental models of diabetes were focused on pancreatectomized (partial or total) animals. Selection of species was more or less spontaneous. Usually used by small animals (mostly rats and mice), both because of handling or space needs and the affordable price. These experiments have often been questioned that the extrapolation to humans is appropriate from experiments on larger animals such as cats, dogs, pigs and primates (McNamara et al., 2009).

Among the non-surgical method inducing hyperglycemia after pancreatic damage with toxic substances, the most famous are streptozotocin (Junod et al., 1969) and alloxan (Lenzen & Panten, 1988; Lenzen, 2008). Surgical and non-surgical methods are a good model for studying the consequences of chronic hyperglycemia and the development of diabetic complications (Salgado et al., 2001). Using female animals it is also possible to study the effects of hyperglycemia to the offspring (Caluwaerts et al., 2003). Problems related to control of hyperglycemia, application of insulin or oral antidiabetic drugs require further improvements of experiments (Herrera et al., 1985). It should be noted that euglycemia cannot be reached in pancreatectomized animals (Harder et al., 2003). In pancreatectomized animals also transplantation of Langerhans' islet cells was studied. Islets cells can be transplanted without capsules or in capsules that protect them from rejection (Lanza et al., 1999), either subcutaneously (Wang et al., 2011), or under the kidney capsule (Korec, 1991, Carlsson et al., 2000) or via the portal vein to the liver (Trimble et al., 1980). Transplanted animals must receive antirejection therapy (Kobayashi et al., 2008). Like in humans is a key event in many successful interventions should be carried out (how many animals to be used) so that we can draw on the basis of their results, conclusions appropriate for transplantation programs.

Rodents are commonly used models for testing the new pharmacologically active substances not only in the context of transplantation (immunosuppressives), but also with regard to therapy or prevention of human diseases. Rats and mice are commonly used in safety and effectiveness testing of new orally active compounds. These experiments also led to unexpected findings about how the PPAR agonists may have a protective effect on the -cells (Bonora et al., 2008). Similarly, in such experiments were also identified the effects of insulin analogues for the development of tumors (Sandow, 2009).

Experimental animals used to study of diabetes mellitus can be essentially classified into several groups (Shafrir, 2003):

- Animals with chemically induced destruction of pancreatic -cells:

  a) Alloxan model

  b) Streptozotocin (single dose)

  c) Streptozotocín (more subdiabetogenic doses).

- Animals with spontaneous autoimmune diabetes:

  a) BB rats

  b) NOD mice

  c) Akita mice

d) LETL rats

e) Torii rats

f) LEW.1AR1/ZTM.

- Genetically altered animals (transgenic models) with various form of diabetes

- Insulin resistant mutants of rodents with potent diabetogenity:

a) C57BKs *db* mice (*lepr*$^{db}$)

b) C57BL6J *ob* mice (*lep*$^{ob}$)

c) Yellow $A^{v}$ a $A^{vy}$ mice

d) KK mice

e) NZO mice

f) Zucker *fa* rats (*lepr*$^{fa}$) and BB/Wor rats

g) Zdf/Drt-*fa* rats

h) Wistar-Kyoto diabetic/fatty rats

i) Obese (corpulent, *cp*) rats of strains SHR/N-*cp*, LA/N-*cp*, SHHF/Mcc-*cp* and JCR.LA-cp.

- Rodents with spontaneous diabetes of various etiology:

a) NON mice

b) WBN/Kob rats

c) ESS rats

d) BHE/Cdb rats

e) OLETF rats

f) NSY mice

g) Koletzky (SHROB) rats ($fa^{K}$)

h) Hypertriglyceridemic (HTG) rats

- Rodents by overfeed induced diabetogenity:

  a) Psammomys obesus (sand rats, gerbil)

  b) Acromys cahirinus (spiny mice)

  c) C57BL/6J mice

- Diabetic rodents obtained by selective crossbreeding from normal strains:

a) GK (Goto-Kakizaki) rats

b) Cohen diabetic rats (induced by diabetogenic diet rich in sucrose and poor in copper).

- Diabetic non-rodents:

  a) Primates

  b) Pigs

  c) Dogs

  d) Cats.

## 3.1. Models Of Spontaneous Type 1 Diabetes Mellitus (T1d)

The most known models of spontaneous T1D include NOD (non-obese diabetic) mice and BB (BioBreeding) rats, having a many common features with human T1D (von Herrath & Nepom, 2009). In addition to these we include here LETL (Long Evans Tokushima Leans) and KDP (Komeda diabetes prone) rats, LEW congenic rats, New Zealand white rabbits, Keeshond dogs (long haired dog of Dutch race), Chinese hamster and Celebes black apes. It should be noted that these animals are kept as inbred in laboratory conditions for many generations and gradually selected to hyperglycemia.

An excellent overview and classification of animal models of T1D is in von Herrath & Nepom (2009):

- Spontaneous or genetically modified diabetic animals
- Non-obese
- NOD mice
- Akita mice
- BB rats
- LETL (Long-Evans Tokushima Lean) rats
- KDP (Komeda diabetes prone) rats
- LEW.1AR1/Ztm-iddm rats
- Monkeys
- Keeshond dogs
- Some races of cats (feline models).
- Chemically induced diabetic animals
- Non-obese
- Single dose of ALX, respectively, more low doses of STZ
- Vacor (rodenticide, which acts as an antagonist of B vitamins, particularly nicotinamide), dithizone, dehydroascorbic acid, pentamidine and 8-hydroxyquinoline.
- Surgically prepared diabetic animals
- Non-obese
- Totally pancreatectomized animals, e.g., dogs, primates, pigs and rats.
- Virus-induced models.

## 3.2. Models Of Spontaneous Type 2 Diabetes Mellitus (T2d) And Monogenic Forms Of Diabetes

T2D is a heterogeneous group of disorders characterized by insulin resistance and impaired insulin secretion, defined by elevated fasting glycemia and hyperglycemia after load of glucose. Some newer subtypes of diabetes are based on single-gene defect, called MODY (Maturity Onset Diabetes of the Young) syndromes (Vaxillaire & Froguel, 2008), syndromes with severe insulin resistance (Semple et al., 2011) and a mitochondrial diabetes (Berdanier, 2007). In most patients, diabetes is caused by several genetic and environmental factors and disease development in all leads to chronic complications.

Animal models of T2D are complex and heterogeneous as in humans (Kaplan & Wagner, 2006). Advances in the interpretation of the problem coming from various sources and models (ob/ob mice – monogenic model of obesity with leptin deficiency, db/db mice – monogenic model of obesity with leptin resistance, Zucker fa/fa rats – monogenic model of obesity with resistance to leptin; Goto Kakizaki rats, KK mice, NSY mice, OLETF rats, Israeli sand [desert] rats, streptozotocin-treated rats receiving fat diet, CBA/Ca mice, New Zealand obese mouse). In some animals, is dominated insulin resistance, compared to other it is damage of pancreatic -cells (Cefalu, 2006). Animals with glucose intolerance and phenotypically more obese with dyslipidemia and hypertension are a good model of human T2DM. Similar to NOD mice and BB rats for T1D, selective inbreeding increases the spontaneous incidence of T2D. Most of the studies come from monogenic models of ob/ob, db/db, fa/fa and agouti strains (Franconi et al., 2008).

Obesity and subsequent insulin resistance are the main triggers of T2D in humans. Due to strong similarities with humans animal models must be obese. Some strains maintain euglycemic status with a strong and sustained compensatory response to -cells, leading to insulin resistance and hyperinsulinemia. Similarly, ob/ob mice and fa/fa rats are a good example of this phenomenon. For others, such as db/db mice and Psammomys obesus which were rapidly develops hyperglycemia and therefore -cells are not able to maintain high levels of insulin. The study of these different animal models may be helpful in explaining why some people with morbid obesity never develop T2D while in others hyperglycemia is already at relatively mild insulin resistance and obesity (Tirabassi et al., 2004).

These models have also contributed significantly to the study of obesity (Srinivasan & Ramarao, 2007). In 1994, Friedman with coworkers cloned ob/ob mice with a mutant gene for severe obesity (Zhang et al., 1994). Normal ob gene encodes a protein secreted by adipocytes and called leptin. In db/db mice and fa/fa rats was found mutations in the gene for the hypothalamic leptin receptor (Lee et al., 1996).

Recent classification of animals for modeling of T2D:

- Spontaneous or genetically modified diabetic animals:
- Obese:
- ob/ob mice
- db/db mice
- KK (Kuo Kondo) mice
- KK/$A^y$ (yellow obese) mice
- NZO mice
- NONcNZO10 mice
- TSOD (Tsumura Suzuki obese diabetic) mice
- M16 mice
- Nagoya-Shibata-Yasuda (NSY) mice
- Zucker fatty rats
- ZDF (Zucker diabetic fatty) rats
- Obese-hyperglycemic Wistar Kyoto rats
- SHR/N-cp rats
- SHHF/Mcc-cp rats
- JCR/LA-cp rats
- OLETF rats
- eSS rats

- BHE/Cdb rats
- Koletzky (SHROB) rats
- Yucatan miniature pigs
- Sinclair miniature pigs
- Göttingen miniature pigs
- Ossabaw pigs
- Familial hypercholesterolinemic pigs (FHP)
- Obese rhesus monkeys
- Macaca fascicularis
- Macaca radiata
- Papo anobis.
- Non-obese:
- WBN/Kob rats
- Goto Kakizaki (GK) rats
- Hypertriglyceridemic (HTG) rats
- Torii rats (SDT, spontaneously diabetic Torii)
- Torii non-obese mice C57BL/6
- ALS/Lt mice
- Non-obese mutant C57BL/6 (Akita) mice.
- With diet/nutrition induced diabetic animals:
- Cohen diabetic rats
- Psammomys obesus

- Acomys cahirinus
- Ctenomys talarum (tuco tuco)
- Guinea pigs (Cricetulus griseus)
- C57/BL 6J mice
- Macaca mullata
- Dogs
- Cats
- Chinese Guizhou mini-pigs.
- Chemically induced diabetic animals:
- Obese:
- GTG (gold thioglucose)–treated obese mice
- Yorkshire and with Yorkshire crossbred strains of pigs (STZ-induced diabetes).
- Non-obese:
- Single low dose of ALX or single dose of STZ to adult rats, mice and etc.
- Neonatal STZ-treated rats.
- Surgically prepared diabetic animals:
- Obese:
- VMH (ventromedial hypothalamus)–damaged dietetically obese diabetic rats.
- Non-obese:
- partially pancreatectomized animals, e.g., dogs, primates, pigs and rats.

- Transgenic/knockout diabetic animals:
- Obese:
- 3 receptor knockout mice
- UCP1 knockout mice.
- Non-obese:
- transgenic or knockout mice in genes for insulin, insulin receptor and their components in the direction of insulin signaling, i.e., IRS-1, IRS-2, GLUT-4, PTP-1B and other
- PPAR- tissue specific knockout mice
- Glukokinase or GLUT-2 gene knockout mice
- HIP rats (rats with overexpession of human islet amyloid polypeptide).

## 3.3. Models Of Spontaneous Type 2 Diabetes Mellitus – Advantages And Disadvatages

In view of advantages and disadvantages of different models of T2D animal models, we can say that:

- Spontaneous or genetically modified diabetic animals have:
- Advantages:
- Development of T2D is a spontaneous and provides genetic factors. In animals developed the similar features as in human T2D.
- Animal models are mostly inbred in which is homogeneous genetic background and environmental factors are well controllable.
- Variability of the results is minimal and requires a small number of animals.
- Disadvantages:

- They are highly inbred, homogeneous, and often with monogenic inheritance therefore development of diabetes is strongly genetically determined versus heterogeneity in humans.
- Limited life span and the time dimension of diabetes study.
- In animals with brittle pancreas (db/db, ZDF rats, P. obesus, and other) is high mortality caused by ketosis and therefore need insulin treatment to prolong their life.
- Require more sophisticated forms of breeding purposes and care.
- With diet/nutrition induced diabetic animals:
- Advantages:
- The development of diabetes associated with obesity is the result of overfeed, as it is also in diabetes in human populations.
- Toxicity of chemical (diabetogenic) compounds to other vital organs and tissues can be eliminated.
- Disadvantages:
- Animal models often require long periods of dietary treatment.
- Not too significant hyperglycemia are after a simple dietary treatment in genetically normal animals and are therefore unsuitable for studying antidiabetogenic substances by determining of blood glucose.
- Chemically induced diabetic animals:
- Advantages:
- Selective loss of -cells of the pancreas (alloxan/STZ), while maintaining of intact - and -cells.
- Residual insulin secretion allows animals to live without insulin therapy a relative long time.
- Ketosis and subsequent mortality is relatively low.
- They are cheaper and easier to maintain.

- Disadvantages:

- Hyperglycemia develops primarily as a result of direct cytotoxic effects on -cells and subsequent insulin deficiency and not as a result of insulin resistance.

- Chemically induced diabetes is less stable and often occurs spontaneously regeneration of pancreatic -cell, which is a disadvantage of long-term studies.

- In chemically induced diabetes can be toxic substance damaged other molecular structures, and may result to the general toxic effects.

- Variability of results with regard of hyperglycemia is very high.

- Surgically prepared diabetic animals:

- Advantages:

- Eliminates the cytotoxic effect of chemical diabetogenic substances to other organs and tissues.

- Strongly resembles regarding reduction of -cell mass to human T2D.

- Disadvantages:

- Technical complexity and cumbersome, postoperative procedures.

- The occurrence of certain digestive problems (as result of excision of the exocrine pancreas, amylase deficiency, etc.).

- Removal of -cells (glucagon producing) together with -cells leads to problems in regulation of hypoglycemic events.

- Mortality is relatively high.

- Transgenic/knockout diabetic animals:

- Advantages:

- The effect of a single gene or their mutations on diabetes can be studied in vivo.

- Greatly facilitate the resolution of the genetic complexity of T2D.

- Disadvantages:
- Highly sophisticated and expensive procedures for production and breeding.
- Very expensive for routine screening tests.

Spontaneous T2D diabetic animals are normally obtained from animals with mutations in one or more genes that are transmitted from generation to generation (e.g., ob/ob, db/db mice) or the selection from non-diabetic outbred animals selectively crosses over several generations (e.g., GK rats, TSOD mice). These animals have congenital diabetes with the monogenic or polygenic defects. Metabolic peculiarities may be due to a single gene defect (monogenic) with a dominant phenotype (e.g., yellow obese KK/Ay mice) and the recessive phenotype (diabetic db/db mice, Zucker fatty rats) or may have polygenic origin (i.e. KK mice, NZO mouse) (Aerts & Van Assche, 2006). T2D in humans is the result of interaction of different environmental factors and many genes which under certain conditions may manifest as diabetes with very contrasting symptomatology (e.g., MODY), single-gene defects with clinically overt diabetes are rare. Therefore, animals with polygenic defects are more objective for a modeling study of T2D in humans (Lofty et al., 2011).

## 4. PROBLEMS OF EVALUATION AND INTERPRETATION OF THE RESULTS FROM ANIMAL EXPERIMENTS

Long-term fluctuations in blood glucose levels can lead to loss of consciousness in animals. Both hyperglycemia and hypoglycemia can lead to diabetic coma, which is characterized by disorientation and convulsions. Diabetic coma is a life-threatening event. Without therapy of diabetic rats, blood glucose levels rising significantly, there is dehydration and electrolyte losses by urine, is not stores the fat and protein, even significantly break down protein and fat reserves. This dysregulation leads to ketosis and the release of ketone bodies in blood. It is unethical not treated animals with glycemia above 25 mM/L. Therefore, certain authors rejected animals from experiment with glycemia above 25 mM/L (Matteucci & Giampietro, 2008).

Similarly, non-treatment of significant hypoglycemia in the initial phase of alloxan diabetes is unethical. Some authors reported that animals are treated by glucose solution to alleviate complications and death due to hypoglycemia during the first 6 hours after administration of alloxan (deCarvalho et al., 2008).

Induction of diabetes after administration of diabetogenic substances is confirmed by the determination of glucose in blood. After blood collection are blood cells separated within one hour with the addition of glycolytic inhibitors. Glucose is then determined in plasma by using of standard enzymatic methods (Sacks et al., 2002). The concentration of glucose in plasma is 11% higher than in whole blood and glucose concentration in heparinized plasma may be 5% lower than in serum. Glucose concentrations during the oral glucose tolerance test in capillary blood are higher (about 20–25%) than in venous blood. Variation coefficient for glucose in plasma is 2.2%. Transmissible glucometers have a much lower sensitivity than the above coefficient of variation. There are similar differences between values measured by different glucometers. American Diabetes Association has built a development goal for blood glucose monitors with analytical deviations 5%. The diagnosis of diabetes in humans is based on the following criteria (ECDCDM, 2003):

- symptoms of diabetes and casual glycemia above 11.1 mM/L;
- fasting glycemia (FPG, Fasting Plasma Glucose) 7.0 mM/L;
- glycemia after glucose load increases over 2 h 11.1 mM/L.

These values can be confirmed the next day by re-examination. It must be noted that these criteria are applicable if the rules for determining the glucose levels are sufficiently accurate and sensitive beyond the range of between 7 and 11.1 mM/L.

When using the blood glucose monitors is needed just a drop of blood (less traumatized animals), although reproducibility and statistical accuracy is lower (needs more tests). As is to harmonize with the principle of the four R?

Oral glucose tolerance test (oGTT) provides information on ability to cope with the load by glucose. In adults, it is used in 3 hours arrangement and applied to 100 g of glucose in the diagnosis of gestational diabetes mellitus, compared with 2 hours arrangement and the application of 75 g glucose was used to confirm a diagnosis of diabetes (prediabetes) in the previous dubious blood glucose levels. In children used 1.75 g glucose/kg of b.w. maximum of 75 g glucose. It is a sensitive test for detection of disorders of glucose metabolism, especially if fasting blood glucose are in the dubious areas. For its correct implementation is necessary to respect a several principles:

- does not reduce the glucose uptake three days before the test;
- test should be performed after fasting overnight (10–14 h);
- 25–30% glucose solution is administered orally;
- glucose levels are measured before and 30, 60 and 120 minutes after administration of glucose;
- during the test do not receive food and water.

In rats given 25–30% aqueous solution of anhydrous glucose at a dose of 1-10 g/kg of b.w. Sampling are quite different before glucose administration and every 30 minutes up to 300 minutes after glucose administration (Matteucci & Giampietro, 2008).

In rats, it would be appropriate to use a similar arrangement of the test than in children with a completed at 120 min, in the case of dubious glucose levels at 8–12 mM/L.

If in rats are found blood glucose levels (between 12–16 mM/L), they can be considered as mild diabetic (or as prediabetic condition). Usually they have after a short period increased blood glucose above 16 mM/L, rats with glycemia above this level are considered as clearly diabetic (Lukačínová et al., 2008).

In the oGTT in rats and humans is difference in glucose load in terms of 1 kg of body weight. These differences are not indicated anything substantial cause.

Intravenous glucose tolerance test (ivGTT) used to determine of insulin secretion and determination of the first phase of insulin response (FPIR, First Phase Insulin Response), which is considered a risk factor for T1D in humans (Culina et al., 2011). When studying the prevention of T1D showed that 2 h oGTT was sensitive about 6 months before diagnosis of diabetes compared with FPIR, which was lower and declined with age. Higher sensitivity was achieved using both tests (Barker et al., 2007). Dextrose was administered at a dose of 0.5 g/kg (maximum 35 g) intravenously over 3 minutes. Blood was taken with the -10, -4, 1, 3, 5, 7 and 10 minutes before and after load with dextrose and analyzed for glucose and insulin. FPIR is expressed on the basis of the sum of values - 1 and 3-minute insulin levels.

In rats, using anhydrous glucose at a dose of 0.001 to 1.0 g/kg of b.w. and blood glucose and insulin are evaluated at different time periods typically from -15 min within 30-120 minutes from the load with glucose. It should be noted that these tests in animals do not have such importance and predictive value as in humans (mainly the system of food intake and fasting before the test). Deprivation of food in animals from the evening before the examination is a powerful stress stimulus (due to their nocturnal activity), which has a significant influence on interindividual differences. Of course it is difficult to determine what dose of load in humans corresponds to a similar stress in rats, what is the possibility of extrapolation of results between these species. Probably there should be a 35 g aliquot of the maximum load in humans (but how and when to realize?). We believe that this model could be accepted to study of impact assessment of preventive or therapeutically active substances.

Intraperitoneal glucose tolerance test (ipGTT) is not used in humans. In rats, the dose of glucose ranging from 0.2 to 2.0 g/kg of b.w. and monitored for three to seven time periods from 0 to 60–120 minutes after load of glucose (Matteucci & Giampietro, 2008). It should therefore be considered only as an experimental model without real impact for extrapolation. This model may be interesting for studies comparing of load by different agents as a possible methodological model.

When studying the problem of diabetes plays an experimental protocol and the possibility of extrapolating a very important role. The experimental protocol must include a detailed description of the methods of research, in an experiment to be included animals suitable for this study and must be chosen a good system of controls, used substances must be best defined and used appropriate methods of statistical analysis of results. E.g., in the case of research in herbal medicine has been submitted WHO guidelines specifying the some principles of experiment (WHO, 1993). The primary aims of non-clinical studies are: (1) determine whether the substance has a beneficial effect in terms of herbal medicine (2) characterize the range of pharmacological effects, (3) define the chemical characteristics of the pharmacologically active natural products and their mechanisms of action.

Pharmacodynamic and current pharmacological investigations used animal models or bioassays which are good models for modeling of human disease. As an experimental object (test system) can be used live animals, isolated organs or tissues, blood and its components, tissue and cell cultures, and various subcellular structures. A very important point is determining the appropriate doses for a given system from viewpoints of study of dose-effect relationship and their extrapolation to human (Resjö et al., 2008). In all studies must be a negative control (vehicle without active substance) and positive control (known drug/substance). The series of examinations should be tests to clarify the biological activity of the herbal preparation. For example, plasma insulin concentrations in relation to blood glucose, liver glycogen and triglyceride levels may help in understanding the absorption and utilization of glucose and the like. Toxicological methods include tests of toxicity (topical, systemic and special). Acute toxicity tests require a sufficient number of dose levels to determine the lethal dose and monitoring should take at least 7–14 days. In chronic tests of toxicity is application period lasts from the 2 weeks to 12 months. Particularly difficult are lifelong toxicity tests (Lukačínová et al., 2011). When rodents are used it is recommended that each group has at least 5–10 animals for both sexes.

In most animal studies, the experimenters assessed the effect of intervention on the basis of the null hypothesis, i.e. assume that the experimental intervention had no effect. Guidelines for construction of animal experiments is why strictly rules necessary for adequate control by using the smallest number of animals (Festing & Altman, 2002). Randomization and the use of blanks are rarer and therefore animal experiments indicate much more positive effects of treatment (Perel et al., 2007). Random choice of animals is essential for selection of animals to experimental (treated) and control groups (usually all individuals are healthy!) and despite does not reflect adequately to human population. Groups are designated by researcher and he knew what was that group treated and has been these groups intervened. In this area, animal experiments will be necessary to objectify (one divided and treated groups, and other these groups evaluated).

Since safety and efficacy of drugs before clinical examination are tested on animals, there are important all efforts to eliminate bias and random errors. Moreover, animal models should be as much as possible related clinical conditions. This is particularly important in extrapolation of dose (g/kg, g/m$^2$ of body surface, resp.). Similarly, when comparing in drug already in use, in animal experiments should be used of dosage as in humans.

# 5. CONCLUSION

The dawn and the subsequent development of experimental medicine in the second half of the XIXth century was unimaginable without the use of animals. Nobody cared about their fate and suffering – they were sacrificed in the war against diseases and for the development of science. The rules of experiment were simple and represented only the needs of the experimentator. The discovery of pancreatic diabetes and the subsequent isolation of insulin are the best examples of this era.

The situation changed dramatically in the second half of the XXth century in the development of new methods on laboratory analysis, in the development of new preventive and therapeutic procedures but especially in the new non-animal and animal models for the study in this area. For this analysis shows that are not yet standardized animal experiments even in the study of diabetes. These differences can lead to different conclusions regarding the pharmacologically active substances used in particular in the prevention but also in treatment of diabetes mellitus, e.g., hypoglycemic effect and dose relations in the application of vanadium or bioflavonoids (Lukačínová et al., 2008). It will be important to present a unified experimental approaches for testing of different substances on animals, both from the aspect of arrangement the experiment (control groups, conditions of experiment, statistical evaluation) as well as the selection of the studied parameters (markers) for individual type of test (acute, chronic, etc.) or the test substance. In this consideration is necessary increased attention to the requirements for the application of 4R. In this case, it can be expected the significant reduction in the need for experimental animals, does not need to repeat certain experiments only because they were not considered some of the basic conditions (often simple parameters such as the weight of the animals, water intake, food intake, urine output, etc.). Therefore, further work is needed to improve and refine existing guidelines for their specific needs for testing of biologically active substances for medical use. Especially at present when more and more come to the forefront of evidence-based medicine (Borgerson, 2005) should become a standard working method in animal experiments.

In any case, in the future is expected many new findings from animal models, particularly in the pathogenesis of human diseases. The immediate benefit of such experiments was the introduction to testing of insulin therapy, as well as testing of other drugs. It is necessary to calculate that there can also lead to no thoroughfares of research, and it should be remembered the reproducibility and extrapolation of results for the human population. We expect the most benefit but in the verification of preventive strategies with various drugs.

## ACKNOWLEDGEMENT

This work was supported by Grant Agency VEGA No. 1/3494/06.

## NOTES

[1] - "Euthanasia" in association with the termination of life in animal experiment is a widely used term which on the other side is a mistake.

## REFERENCES

1. L. Aerts, F. A. Van Assche, 2006Animal evidence for the transgenerational development of diabetes mellitus. Intern. J. Biochem. Cell Biol., 383-4May-June 2006), 8949031357-2725

2. M. A. Atkinson, G. S. Eisenbarth, 2001Type 1 diabetes: New perspectives on disease pathogenesis and treatment. Lancet, 3589277July 2001), 2212290140-6736

3. M. A. Atkinson, E. H. Leiter, 1999The NOD mouse model of insulin dependent diabetes: As good as it gets? Nature Med., 56June 1999), 6016041078-8956

4. R. E. Bark, 1995The fourth R of research. Am. Assoc. Lab. Anim. Sci., 542March 1995), 501559-6109

5. J. M. Barker, K. Mc Fann, L. C. Harrison, S. Fourlanos, J. Krischer, D. Cuthbertson, H. P. Chase, G. S. Eisenbarth, 2007Pre-type 1 diabetes dysmetabolism: maximal sensitivity achieved with both oral and intravenous glucose tolerance testing. J. Pediatrics, 1501January 2007), 31360340-6199

6. C. Benoist, D. Mathis, 1997Autoimmune diabetes: Retrovirus as trigger, precipitator or marker? Nature, 3886645August 1997), 8338340028-0836

7. C. D. Berdanier, 2007Linking mitochondrial function to diabetes mellitus: an animal's tale. Am. J. Physiol. Cell Physiol., 2933September 2007), C830C8360363-6143

8. A. C. Beynen, J. Hau, 2001Animal models. In: Principles of Laboratory Animal Science. van Zutphen, L.F.M., Baumans, V., Beynen, A.C. (Eds): Principles of Laboratory Animal Science. Elsevier, 0-44450-612-8197205

9. E. Blas, Kurup. A. . Sivasankara, Eds, 2010Priority Public Health Conditions: From Learning to Action on Social Determinants of Health. World Health Organization, 978-9-24156-397-0Geneva, 7794

10. M. Bliss, 1982The discovery of insulin. Chicago Univ. Press, 0-22605-897-21304

11. E. Bonora, 2008Protection of pancreatic beta-cells: is it feasible? Nutr. Metab. Cardiovasc. Dis., 181January 2008), 74830939-4753

12. K. Borgerson, 2005Evidence-Based Alternative Medicine? Perspect. Biol. Med., 484Autumn 2005), 5025150031-5982

13. M. Brownlee, 2001Biochemistry and molecular cell biology of diabetic complications. Nature, 4146865December 2001), 81382 00028-0836

14. R. Bucala, Z. Makita, T. Koschinsky, A. Cerami, H. Vlassara, 1993Lipid advanced glycosylation: pathway for lipid oxidation in vivo. Proc. Natl. Acad. Sci. USA, 9014July 1993), 643464380027-8424

15. R. Bucala, H. Vlassara, A. Cerami, 1994Advanced glycosylation end products: role in diabetic and non-diabetic vascular disease. Drug Develop. Res., 322June 1994), 77891098-2299

16. P. Carlsson, O. , F. Palm, A. Andersson, P. Liss, 2000Chronically decreased oxygen tension in rat pancreatic islets transplanted under the kidney capsule. Transplantation, 695March 2000), 7617660041-1337

17. E. J. Calabrese, 1991Principles of Animal Extrapolation: Predicting Human Responses from Animal Studies. Lewis Publishers, Inc., 0-87371-410-5MI., 1616

18. S. Caluwaerts, K. Holenmans, R. Van Bree, J. Verhaeghe, F. A. Van Assche, 2003Is low-dose streptozotocin in rats an adequate model for gestational diabetes mellitus? J. Soc. Gynecol. Investig., 104May 2003), 2162211071-5576

19. V. F. Carvalho, L. V. Campos, F. A. Farias-Filho, L. T. Florim, E. O. Barreto, C. Pirmez, W. Savino, M. A. Martins, P. M. R. Silva, 2008Suppression of allergic inflammatory response in the skin of alloxan-diabetic rats: relationship with reduced local mast cell numbers. Int. Arch. Allergy Immunol., 1473October 2008), 2462541018-2438

20. W. T. Cefalu, 2006Animal models of type 2 diabetes: clinical presentation and pathophysiological relevance to the human condition. ILAR J., 4731861981084-2020

21. S. Culina, C. Boitard, R. Mallone, 2011Antigen-based immune therapeutics for type 1 diabetes: magic bullets or ordinary blanks? Clin. Develop. Immunol.

22. R. B. Dell, S. Hollevan, R. Ramakrishnan, 2002Sample size determination. ILAR J., 4342072131084-2020

23. W. Dong, H. B. Shil, H. , Y. B. Miao, T. J. Liu, W. Wang, 2010Homoisoflavanones from Polygonatum odoratum rhizomes inhibit advanced glycation end product formation. Arch. Pharm. Res., 335May 2010), 6696740253-6269

24. D. G. Dyer, J. A. Blackledge, S. R. Thorpe, J. W. Baynes, 1991Formation of pentosidine during nonenzymatic browning of proteins by glucose. J. Biol. Chem., 26618June 1991), 11654116600021-9258

25. L. Eiselein, H. J. Schwartz, J. C. Rutledge, 2004The challenge of type 1 diabetes mellitus. ILAR J., 4532312361084-2020

26. L. Ettaro, T. J. Songer, P. Zhang, M. M. Engelgau, 2004Cost-of-illness studies in diabetes mellitus. Pharmacoeconomics, 223149 1641170-7690

27. ECDCDM (Expert Committee on the Diagnosis and Classification of Diabetes Mellitus)2003Report of the expert committee on the diagnosis and classification of diabetes mellitus. Diabetes Care, 26Suppl. 1 (January 2003), S5S200149-5992

28. M. F. W. Festing, D. G. Altman, 2002Guidelines for the design and statistical analysis of experiments using laboratory animals. ILAR J., 4342442581084-2020

29. F. Franconi, G. Seghieri, S. Canu, E. Straface, I. Campesi, W. Malorni, 2008Are the available experimental models of type 2 diabetes appropriate for a gender perspective? Pharmacol. Res., 571January 2008). 6181043-6618

30. A. Green, C. C. Patterson, E. U. R. O. D. I. A. B. T. I. G. E. R. S. G. , 2001Trends in the incidence of childhood-onset diabetes in Europe 1989-1998. Diabetologia, 44Suppl. 3 (October 2001), B3B80001-2186X

31. D. L. Greiner, A. A. Rossini, J. P. Mordes, 2001Translating data from animal models into methods for preventing human autoimmune diabetes mellitus: caveat emptor and primum non nocere. Clin. Immunol., 1002August 2001), 1341431521-6616

32. T. Harder, K. Franke, S. Fahrenkrog, L. Aerts, R. Van Bree, F. A. Van Assche, A. Plagemann, 2003Prevention by maternal pancreatic islet transplantation of hypothalamic malformation in offspring of diabetic mother rats is already detectable at weaning. Neurosci. Lett., 3523December 2003), 1631660304-3940

33. J. Hau, 2008Animal models for human disease: an overview. In: Sourcebook of Models for Biomedical Research. Conn, M.P. (Ed.): Humana Press, 978-1-58829-933-8Totowa, New Jersey, 38

34. J. Hau, O. M. Poulsen, 1988Doses for laboratory animals based on metabolic rates. Scand. J. Lab. Anim. Sci., 15281840901-3393

35. E. Herrera, M. Palacin, A. Martin, M. A. Lasuncion, 1985Relationship between maternal and fetal fuels and placental glucose transfer in rats with maternal diabetes of varying severity. Diabetes, 34Suppl. 2, 42460012-1797

36. A. Junod, A. E. Lambert, W. Stauffacher, A. E. Renold, 1969Diabetogenic action of streptozotocin: Relationship of dose to

metabolic response. J. Clin. Invest., 4811November 1969), 212921390000-0021

37. J. R. Kaplan, J. D. Wagner, 2006Type 2 diabetes- an introduction to the development and use of animal models. ILAR J., 4731811851084-2020

38. D. L. Kaufman, M. G. Erlander, M. Clare-Salzler, M. A. Atkinson, N. K. Maclaren, A. J. Tobin, 1992Autoimmunity to two forms of glutamate decarboxylase in insulin-dependent diabetes mellitus. J. Clin. Invest., 891January 1992) 2832920000-0021

39. T. Kobayashi, H. Arefanian, G. Harb, E. B. Tredget, R. V. Rajotte, G. S. Korbutt, G. R. Rayat, 2008Prolonged survival of microencapsulated neonatal porcine islet xenografts in immune-competent mice without antirejection therapy. Cell Transplant., 1711124312560963-6897

40. T. Koppányi, M. A. Avery, 1966Species differences and the clinical trial of new drugs: a review. Clin. Pharmacol. Ther., 72February 1966), 2502700009-9236

41. R. Korec, 1991Experimental and Spontaneous Diabetes Mellitus in the Rat and Mouse. Edition Center of Šafarik University, 8-07097-123-11248

42. A. Kukreja, N. K. Maclaren, 1999Autoimmunity and diabetes. J. Clin. Endocrinol. Metab., 8412December 1999), 437143780002-1972X

43. R. P. Lanza, D. M. Ecker, W. M. Kühtreiber, J. P. Marsh, J. Ringeling, W. L. Chick, 1999Transplantation of islets using microencapsulation: studies in diabetic rodents and dogs. J. Molecular Medicine, 771January 1999), 2062100946-2716

44. G. H. Lee, R. Proenca, J. M. Montez, K. M. Carroll, J. G. Darvishzadeh, J. I. Lee, J. M. Friedman, 1996Abnormal splicing of the leptin receptor in diabetic mice. Nature, 3796566February 1996), 6326350028-0836

45. E. H. Leiter, 1998NOD mice and related strains: Origins, husbandry, and biology. In: NOD Mice and Related Strains: Research Applications in Diabetes, AIDS, Cancer and Other Diseases. Leiter,

E.H., Atkinson, M.A. (Eds.): Internet: R.G. Landes Co., ISBN 157059466X, 136

46. E. H. Leiter, Herrath. M. von, 2004Animal models have little to teach us about type 1 diabetes: 2. In opposition to this proposal. Diabetologia, 4710October 2004), 165716600001-2186X

47. S. Lenzen, 2008The mechanisms of alloxan- and streptozotocin-induced diabetes. Diabetologia, 512February 2008), 2162260001-2186X

48. S. Lenzen, U. Panten, 1988Alloxan: history and mechanism of action. Diabetologia, 316June 1988), 3373420001-2186X

49. F. M. Loew, 1996Using animals in research. In: Birth to Death: Science and Bioethics. Thomasma, D.C., Kushner, T. (Eds): Cambridge University Press, 0-52155-556-6301312

50. M. Lotfy, J. Singh, H. Kalász, K. Tekes, E. Adeghate, 2011Medicinal chemistry and applications of incretins and DPP-4 inhibitors in the treatment of type 2 diabetes mellitus. Open Med. Chem. J., 5Supple 2-M4 (9 September 2011), 82921874-1045

51. A. Lukačínová, J. Mojžiš, R. Beňačka, J. Keller, T. Maguth, P. Kurila, L. Vaško, O. Rácz, F. Ništiar, 2008Preventive effects of flavonoids on alloxan-induced diabetes mellitus in rats. Acta Vet. Brno, 772June 2008), 1751820001-7213

52. A. Lukačínová, O. Rácz, E. Lovásová, F. Ništiar, 2011Effect of lifetime low dose exposure to heavy metals on selected serum proteins of Wistar rats during three subsequent generations. Ecotoxicol. Environ. Saf., 746September 2011), 174717550147-6513

53. R. D. Mann, 1984Modern Drug Use; An Enquiry on Historical Principles. MTP Press Ltd., 0-85200-717-5and London, 1769

54. E. Matteucci, O. Giampietro, 2008Proposal open for discussion: defining agreed diagnostic procedures in experimental diabetes research. J. Ethnopharm., 1152January 2008), 1631720378-8741

55. J. M. Mc Namara, P. A. Stephens, S. R. X. Dall, A. I. Houston, 2009Evolution of trust and trustworthiness: social awareness favours personality differences. Proc. R. Soc. B, 2761657February 2009), 6056131471-2954

56. M. T. Miller, K. Strömland, 1999Teratogen update: Thalidomide: a review, with a focus on ocular findings and now potentional uses. Teratology, 605November 1999), 3063211542-0760

57. O. Minkowski, 1989Historical development of the theory of pancreatic diabetes. Diabetes, 381January 1989), 160012-1797

58. J. P. Mordes, R. Bortell, H. Groen, D. L. Guberski, A. A. Rossini, D. L. Greiner, 2001Autoimmune diabetes mellitus in the BB rat. In: Animal Models of Diabetes: A Primer. Sima, A.A.F., Shafrir, E. (Eds.). Harwood Academic Publishers, 9-05823-096-1141

59. F. Ništiar, A. Ništiarová, R. Beňačka, O. Rácz, A. Lukačínová, 2006Quo vadis experiments on animals in the university education and tesearch? Slovak Veter. J., 3152892911335-0099

60. N. Ohgami, R. Nagai, M. Ikemoto, H. Arai, A. Kuniyasu, S. Horiuchi, H. Nakayama, 2001Cd36, a member of the class b scavenger receptor family, as a receptor for advanced glycation end products. J. Biol. Chem., 2762February 2001), 319532020021-9258

61. H. Olson, G. Betton, D. Robinson, K. Thomas, A. Monro, G. Kolaja, P. Lilly, J. Sanders, G. Sipes, W. Bracken, M. Dorato, K. Van Deun, P. Smith, B. Berger, A. Heller, 2000Concordance of the toxicity of pharmaceuticals in humans and in animals. Regul. Toxicol. Pharmacol., 321August 2000), 56670273-2300

62. M. Peppa, J. Uribarri, H. Vlassara, 2004The role of advanced glycation end products in the development of atherosclerosis. Curr. Diabetes Rep., 41February 2004), 31361534-4827

63. P. Perel, I. Roberts, E. Sena, P. Wheble, C. Briscoe, P. Sandercock, M. Macleod, L. E. Mignini, P. Jayaram, K. S. Khan, 2007Comparison of treatment effects between animal experiments and clinical trials: systematic review. BMJ, 3347586January 2007), 1972020959-8138

64. P. Perry, 2007The ethics of animal research: a UK perspective. ILAR J., 48142461084-2020

65. P. Quehenberger, A. Bierhaus, P. Fasching, C. Muellner, M. Klevesath, M. Hong, G. Stier, M. Sattler, E. Schleicher, W. Speiser, P. P. Nawroth, 2000Endothelin-1 transcription is controlled by

nuclear factor-kappa B in AGE-stimulated cultured endothelial cells. Diabetes, 499September 2000), 156115700012-1797

66. D. A. Rees, J. C. Alcolado, 2005Animal models of diabetes mellitus. Diabetic Medicine, 224April 2005), 3593701464-5491

67. R. Resjö, K. Berger, M. Fex, O. Hansson, 2008Proteomic studies in animal models of diabetes. Proteomics Clin. Appl., 25May 2008), 6546691862-8354

68. G. Roglic, N. Unwin, 2010Mortality attributable to diabetes: estimates for the year 2010. Diabetes Res. Clin. Pract., 871January 2010), 15190168-8227

69. A. A. Rossini, M. J. Thompson, J. P. Mordes, 2003Diabetes mellitus. In: Office Practice of Medicine. 4th ed., Branch, W.T. Jr. (Ed.): Saunders, 0-72167-672-3695740

70. W. M. S. Russell, R. L. Burch, 1959The Principles of Humane Experimental Technique. Methuen & Co. Ltd., 0-90076-778-21238

71. D. B. Sacks, D. E. Bruns, D. E. Goldstein, N. K. Maclaren, J. M. Mc Donald, M. Parrott, 2002Guidelines and recommendations for laboratory analysis in the diagnosis and management of diabetes mellitus. Clin. Chem., 483March 2002), 4364721434-6621

72. H. C. Salgado, R. Fazan, V. P. S. Fazan, da. Dias, V. J. Silva, A. A. Barreira, 2001Arterial baroreceptors and experimental diabetes. Ann. N. Y. Acad. Sci., 9401September 2001), 20271749-6632

73. J. Sandow, 2009Growth effects of insulin and insulin analogues. Arch. Physiol. Biochem., 1152May 2009), 72851744-4160

74. R. K. Semple, D. B. Savage, E. K. Cochran, P. Gorden, S. O'Rahilly, 2011Genetic syndromes of severe insulin resistance. Endocr. Rev., 324August 2011), 4985140016-3769X

75. D. V. Serreze, E. H. Leiter, 2001Genes and pathways underlying autoimmune diabetes in NOD mice. In: Molecular Pathology of Insulin Dependent Diabetes Mellitus. Von Herrath M.G. (Ed.): Karger Press, 3-80557-240-93167

76. E. Shafrir, 2003Diabetes in animals: contribution to the understanding of diabetes by study of its etiopathology in animal

models. In: Diabetes Mellitus, Porte, D., Sherwin, R.S., Baron, A. (Eds.), McGraw-Hill, 0-83852-041-3York, 231255

77. J. Shaw, R. Sicree, P. Zimmet, 2010Global estimates of the prevalence of diabetes for 2010 and 2030. Diabetes Res. Clin. Pract., 87, 414

78. A. A. F. Sima, E. . Shafrir, Eds, 2000Animal Models of Diabetes: A Primer. Harwood Academic Publishers, 9-05823-096-11364

79. J. S. Skyler, D. Brown, H. P. Chase, E. Collier, C. Cowie, G. S. Eisenbarth, J. Fradkin, G. Grave, C. Greenbaum, R. A. Jackson, F. R. Kaufman, J. P. Krischer, J. B. Marks, J. P. Palmer, A. Ricker, D. A. Schatz, D. Wilson, W. E. Winter, J. Wolfsdorf, A. Zeidler, H. Dickler, R. C. Eastman, N. K. Maclaren, J. I. Malone, 2002Effects of insulin in relatives of patients with type 1 diabetes mellitus. N. Engl. J. Med., 34622May 2002). 1685B1691B0028-4793

80. K. Srinivasan, P. Ramarao, 2007Animal models in type 2 diabetes research: an overview. Indian J. Med. Res., 1253March 2007), 4514720971-5916

81. D. M. Stern, S. D. Yan, S. F. Yan, A. M. Schmidt, 2002Receptor for advanced glycation end products (RAGE) and the complications for diabetes. Ageing Res. Rev., 11February 2002), 1151568-1637

82. W. L. Suarez-Pinzon, A. Rabinovitch, 2001Approaches to type 1 diabetes prevention by intervention in cytokine immunoregulatory circuits. Int. J. Exp. Diabetes Res., 213171560-4284

83. G. H. Tesch, D. J. Nikolic-Paterson, 2006Recent Insights into experimental mouse models of diabetic nephropathy, The Nephron Journals, 1042September 2006), 57621660-8151

84. R. S. Tirabassi, J. F. Flanagan, T. Wu, E. H. Kislauskis, P. J. Birckbichler, D. L. Guberski, 2004The BBZDR/Wor rat model for investigating the complications of type 2 diabetes mellitus. ILAR J., 4532923021084-2020

85. E. R. Trimble, E. G. Siegel, H. Berthoud, R. , A. E. Renold, 1980Intraportal islet transplantation: functional assessment in conscious unrestrained rats. Endocrinology, 1063March 1980), 7917970013-7227

86. N. Van der Werf, F. G. Kroese, J. Rozing, J. L. Hillebrands, 2007Viral infections as potential triggers of type 1 diabetes. Diabetes Metab. Res. Rev., 233March/April 2007), 1691831520-7552

87. M. Vaxillaire, P. Froguel, 2008Monogenic diabetes in the young, pharmacogenetics and relevance to multifactorial forms of Type 2 diabetes. Endocr. Rev., 293May 2008), 2542640016-3769X

88. H. Vlassara, 1997Recent progress in advanced glycation end products and diabetic complications. Diabetes, 46Suppl. 2, S19S250012-1797

89. Herrath. M. Von, G. T. Nepom, 2009Animal models of human type 1 diabetes. Nature Immunol., 102February 2009), 1291321529-2908

90. R. Wada, S. Yagihashi, 2005Role of advanced glycation end products and their receptors in development of diabetic neuropathy. Ann. N. Y. Acad. Sci., 10431June 2005), 5986041749-6632

91. P. Wang, M. V. Yigit, Z. Medarova, L. Wei, G. Dai, C. Schuetz, A. Moore, 2011Combined small interfering RNA therapy and in vivo magnetic resonance imaging in islet transplantation. Diabetes, 602February 2011), 5655710012-1797

92. T. Wendt, L. Bucciarelli, W. Qu, Y. Lu, S. F. Yan, D. M. Stern, A. M. Schmidt, 2002Receptor for advanced glycation endproducts (RAGE) and vascular inflammation: Insights into the pathogenesis of macrovascular complications in diabetes. Curr. Atheroscler. Rep., 43May 2002), 2282371523-3804

93. D. Whiting, N. Unwin, G. Roglic, 2010Diabetes: equity and social determinants. In: Priority Public Health Conditions: From Learning to Action on Social Determinants of Health. Blas, E., Sivasankara Kurup A. (Eds.). World Health Organization, 978-9-24156-397-0Geneva, 7794

94. WHO (World Health Organization)1993Research Guidelines for Evaluating the Safety and Efficacy of Herbal Medicines. Regional Office for the Western Pacific, 9-29061-110-3194

95. S. Wild, G. Roglic, A. Green, R. Sicree, H. King, 2004Global prevalence of diabetes: estimates for the year 2000 and projections for 2030. Diabetes Care, 275May 2994), 104710530149-5992

96. G. Williams, J. C. Pickup, 2004Handbook of Diabetes. 3rd ed., Blackwell Publishing Ltd., 1-40512-052-5MA, 1288

97. Y. Yang, P. Santamaria, 2006Lessons of autoimmune diabetes from animal models. Clinic. Sci., 1106June 2006), 6276390143-5221

98. Y. Zhang, R. Proenca, M. Maffei, M. Barone, L. Leopold, J. M. Friedman, 1994Positional cloning of the mouse obese gene and its human homologue. Nature (Lond.), 3726505December 1994), 4254320028-0836

# CHAPTER 7

# Selenium Biochemistry and Bioavailability: Implications for Animal Agriculture

**Shaniko Shini [1,*], Asad Sultan [2] and Wayne L. Bryden [3]**

[1]School of Biomedical Sciences, University of Queensland, St Lucia, QLD 4072, Australia

[2]Department of Poultry Science, Khyber Pakhtunkhwa Agricultural University, Peshawar 25120, Pakistan

[3]School of Agriculture and Food Sciences, University of Queensland, Gatton, QLD 4343, Australia

## ABSTRACT

Selenium (Se) is an essential trace mineral required for growth, development, immune function, and metabolism. Selenium exerts its biological effects as an integral component of selenoproteins (SePs). Deficiency or low Se status leads to marked changes in many biochemical pathways and a range of pathologies and disorders which are associated with SeP function. Animals, and presumably humans, are able to efficiently utilize nutritionally adequate levels of Se in both organic and inorganic forms. It is now clear that the bioavailability of Se varies depending on the source and chemical form of the Se supplement. There are a range of products available for dietary Se supplementation, however, organic sources have been shown to be assimilated more efficiently than inorganic compounds and are considered to be less toxic and more appropriate as a feed supplement. Yeast enriched with

Selenohomoalanthionine (SeHLan) has recently become commercially available, and initial research suggests that it may be an efficacious source for the production of Se enriched animal products.

## KEYWORDS

selenium; selenoproteins; biochemistry; bioavailability; deficiency; excess; requirements; supplements; organic compounds

## 1. INTRODUCTION

Selenium (Se) is an essential trace element for animals and humans. It was discovered in 1818 and named Selene after the Greek goddess of the moon. However, it was not until the 1950s that it was recognized as an essential dietary micronutrient for mammals [1]. Selenium presents a nutritional conundrum because it is an essential trace element with a narrow range between dietary adequacy and toxicity. It is this characteristic of Se that allows it to be considered as either a friend, or, a foe.

Selenium exerts its biological effects as an integral component of selenoproteins (SePs) that contain selenocysteine at their active site [2]. Some thirty SePs, mostly enzymes, have been identified, including a series of glutathione peroxidases, thioredoxin reductases, and iodothyronine deiodinases [1,2,3,4]. The majority play important roles in redox regulation, detoxification, immunity and viral suppression [5,6,7]. Deficiency or low Se status leads to marked changes in many biochemical pathways and a range of pathologies associated with defects of SeP function may occur [8,9,10].

Selenium content of soils can vary widely. In areas where soils are low in bioavailable Se, deficiencies can occur in humans and animals consuming plant based foods grown in those soils. Selenium deficiency have been reported in many countries including China, Japan, Korea, and Siberia, Northern Europe, USA, Canada, New Zealand, and Australia [11,12,13]. Within each country there are large regional differences in soil Se status and in some localities there are plants that accumulate Se resulting in Se toxicity (or selenosis) in grazing animals. However, health outcomes are not only dependent on the total Se content of soils, but also on the amount of Se taken up into plants and animals—the bioavailable selenium [11].

Dietary Se supplementation was first permitted some forty years ago [13]. Since then, there have been significant advances in our knowledge of Se metabolism and the important role that Se plays in animal productivity and health [3,6,9]. During this period, Se has become an important addition to dietary supplements for animals [14]. The following pages provide an overview of the nutritional biochemistry of Se, and an update on recent developments in Se bioavailability and the possible impact on Se enrichment of animal products.

## 2. BIOCHEMISTRY OF SELENIUM AND SELENOPROTEINS

Selenium exists in four oxidation states: elemental Se ($Se^0$), selenide ($Se^{-2}$), selenite ($Se^{+4}$), and selenate ($Se^{+6}$) in a variety of inorganic and organic matrices [14]. The natural soluble inorganic forms, selenite and selenate, account for the majority of total global selenium [12]. The ultimate source of all Se is the rocks and soils of our terrestrial environment. Most soils contain from 0.1 to 2 μg/g of Se, but this is not evenly distributed. Some areas of the world have soils rich in Se (100 μg/g), while many areas have deficient soils (none or >0.1 μg/g) [11]. Organically bound selenide compounds are predominantly seleno-amino acids; the principle chemical form of Se in animal tissues is selenocysteine, while selenomethionine predominates in plants [15].

The chemistry of selenium resembles that of sulphur in several respects but these elements are not completely interchangeable in animal systems. Both, sulphur and Se occur in proteins as constituents of amino acids. Sulphur is one of the most prevalent elements in the body and is present in the sulphur-containing amino acids: methionine, cysteine, homocysteine and taurine. Selenium is a trace element and a component of the amino acids selenocysteine, and selenomethionine [2]. Selenocysteine is identical to cysteine except that sulphur is replaced by a Se atom, which is typically ionized at physiological pH when compared with sulphur-containing species.

Selenocysteine (from animal tissues) and selenomethionine (from plants, marine algae, yeast and bacteria) are both sources of Se suitable for synthesis of SePs [16]. Selenocysteine is incorporated into selenium-containing proteins, designated selenoproteins, following the action of a specific codon for selenocysteine residue [3]. Replacement of selenocysteine by cysteine in a SeP usually results in a dramatic decrease of enzymatic activity, confirming that the ionized Se atom is critical for optimum protein function [9]. Significantly, within all cell types there is a specific biosynthesis pathway that facilitates selenocysteine synthesis and its subsequent incorporation into SePs. Cellular Se concentrations are therefore tightly regulated. The regulation of SeP synthesis is central to understanding Se homeostasis and disorders following the failure of homeostasis [5,17,18]. Cellular Se concentration is a key regulator of Se incorporation into SePs and acts mainly at the post-transcriptional level in response to alterations in Se availability [19]. Selenocysteine biosynthesis represents the main regulatory point for SeP synthesis and not absorption as occurs with many nutrients. In contrast, ingested selenomethionine can be nonspecifically incorporated into proteins in place of methionine or converted to selenocysteine by the trans-sulphuration pathway in liver or kidney and may be used as a biological Se pool [6,17].

The biochemistry of Se is different from most other trace elements as it is incorporated in proteins (SePs) at their highest level of complexity and function. Selenoproteins incorporate selenium only in the form of selenocysteine and this occurs during translation in the ribosome using a transfer RNA specific for selenocysteine. Seleno-amino acids (selenocysteine or selenocystine and selenomethionine) are required for the synthesis of selenium-containing peptides and proteins [15]. There are no known human or animal functionally active SePs that contain selenomethionine.

Only proteins that are genetically programmed and perform essential biological functions are classified as SePs [3]. Some of these SePs are enzymes such as the six antioxidant glutathione peroxidases and the three thioredoxin reductases; the three deiodinases are involved in thyroid function by catalyzing the activation and deactivation of the thyroid hormones. [20]. Selenoprotein P, for example, functions as a transporter of selenium between the liver and other organs [4]. The functional characterization of many SePs remains to be delineated.

# 3. SELENIUM ABSORPTION, DISTRIBUTION AND METABOLIC FATE

An overview of the metabolism of Se is shown in Figure 1. Absorption of selenium occurs in the small intestine, where both inorganic and organic forms of Se are readily absorbed. Selenite is passively absorbed across the gut wall, while selenate appears to be transported by a sodium-mediated carrier mechanism shared with sulphur [14,21]. Organic forms of Se are actively transported. Selenium is distributed throughout the body from the liver to the brain, pancreas and kidneys. The highest Se concentrations are found in the liver and kidneys but the greatest total concentration occurs in muscle because of their proportion of body weight [22]. However, dose–response studies conducted with chicken and lamb have found that in muscle tissue, the proportion of selenomethionine and selenocysteine was equivalent [23], suggesting that both selenomethionine and selenocysteine-containing proteins are able to be absorbed and transferred to peripheral tissues. However, the mechanism is unclear. Selenium is transported in blood by two SePs; selenoprotein P (SePP1) and extracellular glutathione peroxidase (GSH-Px) [6]. Plasma SePP1 contains more than 50% of circulating Se [6,19]. Only insignificant transitory amounts of free selenomethionine are found in blood. It has also been shown, that dietary supplementation with Se enriched yeast increased blood selenomethionine in a dose dependent manner whereas selenocysteine remained constant [24]. These changes in circulating selenised amino acids may reflect saturation of selenoenzymes and a shift from incorporation of Se into functional SePs to non-specific incorporation of selenomethionine into liver tissue protein. This may indicate that alternative transport mechanisms for selenomethionine exist and need to be clarified through further research. Following protein turnover, the released Se, can be recycled via enterohepatic circulation or excreted. Selenium is eliminated primarily in urine and faeces [17]. The distribution between the two routes varies with the level of exposure and time after exposure.

In ruminants, selenite is the primary compound available for absorption because the reducing conditions within the rumen convert the majority of selenate to selenite [26]. In the rumen, about a third of selenite is converted to insoluble forms that are passed into manure. Of the soluble selenite that reaches the intestine, some 40% will be absorbed, compared to about 80% of selenomethionine [27]. Subsequently, in ruminants, the digestibility of Se from selenite is around 50% compared to about 66% for selenium-yeast [27]. There is no information on the impact of the gut microbiota on the Se requirements of monogastric animals.

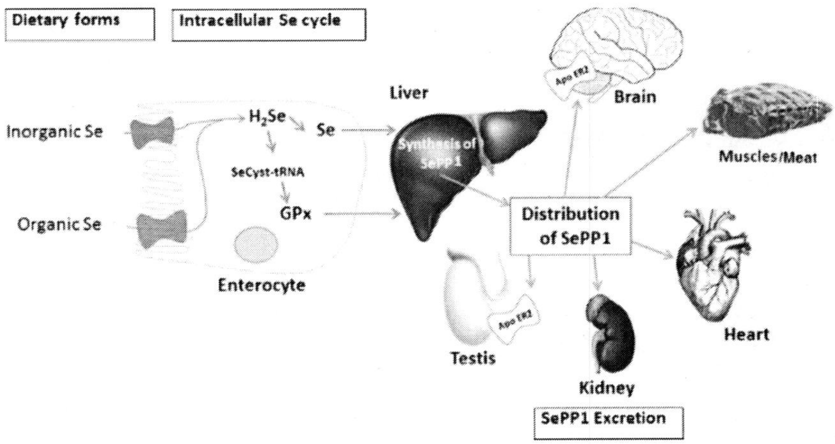


**Figure 1.** General pathways of Se absorption, hepatic synthesis of SePP1 and distribution to various organs. Adapted from Kumar and Priyadarsini [25].

Inorganic Se is recognized by the digestive tissues and is absorbed and converted into SePs [27]. In contrast, organic Se (selenomethionine) is not recognized as Se-containing by mammalian cells [28]. Hence, selenomethionine is absorbed and metabolized relative to methionine needs. If selenomethionine is broken down within the cell, Se is released and recognized by the cell as a mineral [29]. It is then processed according to the need for Se. However, if the cell does not break down selenomethionine, it may be inadvertently incorporated into a wide variety of proteins that are not genetically programmed to contain Se [30], rather than be excreted rapidly. Such incorrect incorporation of Se into protein may be toxic [3,31]. As a metabolic safeguard, neither dietary selenocysteine nor selenomethionine is directly incorporated into selenoproteins. All dietary forms of selenium must be metabolized and converted to selenocysteine and selenoproteins under the genetically controlled mechanism within the cell [32]. Much of the absorbed organic Se is transferred into the amino acid pool, where together with the existing intracellular pool it is metabolized by different pathways (Figure 1). From there, it is enzymatically converted in the liver to selenide, which serves as the Se source for selenocysteine synthesis [33].

# 4. SELENIUM DEFICIENCY AND REQUIREMENTS

Selenium acts biochemically in mammals and birds in a complimentary manner to vitamin E [34,35,36]. Both nutrients prevent peroxidation of unsaturated fatty acids in cell membranes. Most of the deficiency signs of these nutrients can be explained by their antioxidant properties. The requirement for each is therefore influenced by the dietary concentration of the other [2,37]. For example, the Se requirement of the chick is inversely proportional to dietary vitamin E intake. Thus Se has sparing effect on the requirement for vitamin E and vice versa; Se-dependent peroxidases essentially serve as antioxidants when vitamin E levels are not sufficient to prevent lipid peroxidation. On the other hand, vitamin E neutralizes oxidants before the initiation of chain reactions thus reducing Se loss from the body [16]. The metabolic interrelationship of GSH-Px with vitamin E is particularly evident in deficiency diseases that can be prevented either by vitamin E or Se [1,37]. The other major role for Se is the production of thyroid hormones as a component of the enzyme type 1 iodothyronine deiodinase which converts triiodothyronine to thyroxine [1].

Manifestation of Se deficiency can take many forms and varies between species. Muscular degeneration or white muscle disease occurs to varying degrees in all species [14,38]. In birds, pancreatic fibrosis is an uncomplicated Se deficiency, whereas exudative diathesis (generalised oedema visible under the skin) is responsive to both Se and vitamin E [34]. Pigs with hepatosis diatetica (severe necrotic liver lesions) are responsive to Se supplements [39], while both Se and vitamin E are effective in treating mulberry heart disease (a dietetic microangiopathy) [40]. Reproductive disorders, including retained placenta in dairy cows [41], while lowered disease resistance are observed in all Se deficient species [26,27]. Some species, such as rabbits and horses, seem to be more dependent on vitamin E than Se for their antioxidant protection [42]. This may reflect species differences in dependence on non-selenium containing GSH-Px.

Selenium presents a nutritional conundrum because it is both essential and highly toxic. Adequate Se intake levels have a relatively narrow range between deficiency and toxicity. Dietary Se levels of 0.2–0.3 mg Se/kg diet have shown to prevent deficiency and resultant diseases such as white muscle disease in cattle and sheep, exertional myopathy in horses, hepatosis dietetica in pigs, and exudative diathesis in chickens [22,37]. The tolerable concentrations for Se in most livestock feed is considered to be 2–5 mg Se/kg diet, although some believe 4–5 mg Se/kg diet can inhibit growth. Animals that consume diets containing 5–40 mg Se/kg diet over a period of several weeks or months suffer from chronic selenosis [37].

There are several approaches to measuring Se status. These include the measurement of changes in plasma Se concentration, measurement of GSH-Px enzyme activity, and absorption/retention studies. The use of stable isotopes of Se has been used in human studies and to determine endogenous forms of selenium in foods [43]. All of these biomarkers are useful indicators of Se status but because of the role of Se in many biochemical pathways, a single indicator may not be an appropriate index of Se status.

## 5. DIETARY SELENIUM SUPPLEMENTATION

Soils are the major source of Se for plants, while plants are major sources of Se for animals. To improve Se status of plants grown on deficient soils, Se enriched fertilizers may be used or plant breeding strategies may be applied [13]. Despite these strategies, Se is also routinely added to animal diets to ensure that requirements are met. There has been increased interest recently in dietary supplementation with Se to enrich animal products. The production of selenium-enriched meat, milk and eggs is viewed as an effective and safe way of improving the selenium status of humans [44].

There are a range of products available for dietary Se supplementation (Table 1). Selenium is commonly added to diets as sodium selenite. However, there has been growing interest in dietary addition of organic Se. Organic sources are assimilated more efficiently than inorganic Se and considered to be less toxic and therefore more appropriate as a feed supplement [22,45]. Yeast has become the most popular vehicle for the addition of organic Se because of its rapid growth, ease of culture and high capacity to accumulate Se [46,47,48]. The major product in selenized yeast is selenomethionine.

Selenomethionine was found to be four times more effective than selenite in preventing the characteristic pancreatic degeneration caused by selenium deficiency in chicks [49]. Selenium yeast (selenomethionine) was found to be much more effective than inorganic Se in increasing the Se concentration of cow's milk [31]. This is in accord with many animal studies and human clinical trials that have demonstrated the superior efficacy of L-selenomethionine, in increasing Se muscle content compared to inorganic Se [45,50].

Selenohomolanthionine (SeHLan; 2 hydroxy-4-methylselenobutanoic acid) was recently identified in Japanese pungent radish and has generated much interest as it was less toxic in human cell culture than selenomethionine [53,56]. As shown in Figure 2, differences in metabolism between SeHLan and selenomethionine may, in-part, explain the apparent difference in toxicity. Selenomethionine mimics methionine by sharing the same metabolic pathways and can replace methionine in peptide synthesis, and as noted above, thus causing signs of selenosis [31]. The proposed metabolic pathway for SeHLan (Figure 2) appears to be much less complex; SeHLan is only utilized in the trans-selenation pathway for selenoprotein synthesis and therefore is not expected to interfere with the methionine metabolic pathways. The tissue distribution of these two selenoamino acids may also contribute to differences in toxicity [6]. Both are distributed throughout the body with higher liver and pancreas accumulation of selenomethionine in contrast to SeHLan which preferentially accumulates in the liver and kidneys [19]. At higher doses, selenomethionine has been shown to induce pancreas damage [55,57] whereas SeHLan is excreted by the kidneys without inducing pancreatic damage [53]. Tsuji *et al*., have reported specific toxicity of SeHLan to the kidneys which could be avoided when it is administered at a lower dose than in their study [53].

Selenomethionine enriched yeast has been used as a feed supplement in animal agriculture for many years. Recently, a yeast product enriched with SeHLan has become available and efficacy studies with growing pigs [58] and broiler chickens [54] have been conducted with these selenoamino acid sources. In the studies both selenomethionine (Sel Plex) and SeHLan (AB Tor-Sel) were compared to sodium selenite. In clean research, experimental conditions, as demonstrated on many occasions, dietary supplementation with both inorganic and organic selenium resulted in similar animal and bird performance. However, tissue accumulation was significantly greater when the organic forms of Se were fed, which is in accord with the literature [31,48,49,50]. Interestingly, the yeast enriched with SeHLan generated significantly higher Se concentrations in muscle tissue of broiler chickens than the selenomethionine enriched product [54]. The implication of this finding in both pig and broilers may imply a greater efficacy of SeHLan in stressful commercial environments. These interesting results should be the subject of further research.

**Table 1.** Selenium compounds and their application in animal and human nutrition.

| Name and Content | Nature or Origin | Uses | Reference * |
|---|---|---|---|
| Sodium selenate<br>Sodium selenite<br>Selenase 50 mcg/mL | Synthetic Inorganic | For short-term selenium supplementation; orally in the diet, or by injections for both animals and humans | [12,36,44,45,49,50] |
| Sintomin BIOSEL 2000 | Inactive dry yeast containing high levels of organic selenium | All animals | [51] |

| | | | |
|---|---|---|---|
| Selyeast Selenomethionine: 1000, 2000, 3000 | Yeast rich in organic selenium | For use as animal feed. | [22,44,45, 47,48] |
| Selemax® (1000, 2000) 70% of total selenium in the form of seleno-methionine | Inactive dry yeast containing organic selenium | All animal species and categories | [22,45,52] |
| SeLECT L(+) Selenomethionine & Vitamin E | Organic, pure selenomethionine | Oral administration humans (capsules), and animals | [22,44, 45,49] |
| Sel Plex ™ > 50% of total selenium in the form of seleno-methionine | Yeast rich in organic selenium | All animals | [22–24,44, 45,50,51] |
| BiOnyc® Tor-Sel Selenohomolanthionine (SeHLan) | Yeast rich in organic selenium | All animals and humans | [53,54] |
| L-Selenomethionine 100% L-selenomethionine | Naturally occurring organoselenium compound made by plants | Predominate form of selenium supplement in food for humans; some use in animals | [22,44, 45,53] |
| SeMC ™ Methylselenocysteine 98% | Naturally occurring organoselenium compound made by plants | Humans | [22,53,55] |

* Source: generated from references and product information sheets.

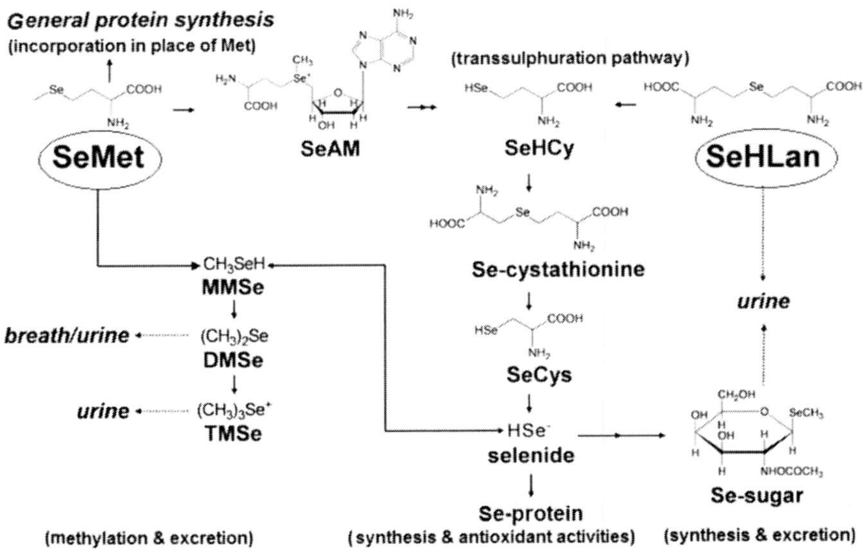


**Figure 2.** Proposed metabolic pathways for SeHLan and SeMet in animal cells. Adapted from Tsuji *et al*. 2010 [53].

# 6. CONCLUSIONS

Selenium's nutritional essentiality was discovered in the 1950s. It is now clear that the importance of having adequate amounts of Se in the diet is primarily due to the fact that this micronutrient is required for the biosynthesis of selenocysteine as a part of functional selenoproteins. Improved analytical techniques will be required to answer those questions that still remain regarding Se metabolism. Although animals, and presumably humans, are able to efficiently utilize nutritionally adequate levels of Se in both organic and inorganic forms for SeP synthesis, it is clear that the bioavailability of Se varies, depending on the source and chemical form of the Se supplement. Tissue enrichment with Se is greater when organic forms of the micronutrient are fed. Greater tissue reserves of Se may enhance an animal's resilience to stress and disease challenge.

Organic Se, in the form of yeast enriched with selenomethionine, is widely used in animal nutrition. Recently, yeast enriched with SeHLan became commercially available and initial research suggests that it may also be an efficacious source for tissue accumulation of Se. This has obvious implications for the production of Se enriched animal products.

## CONFLICTS OF INTEREST

The author declares no conflict of interest.

## REFERENCES


1. Brown, K.M.; Arthur, J.R. Selenium, selenoproteins and human health: A review. *Public Health Nutr.* **2001**, *4*, 593–599.

2. Bock, A.; Forchhammer, K.; Heider, J.; Leinfelder, W.; Sawers, G.; Veprek, B.; Zinoni, F. Selenocysteine: The 21st amino acid. *Mol. Microbiol.* **1991**, *5*, 515–520.

3. Hatfield, D.L.; Gladyshev, V.N. How Selenium Has Altered Our Understanding of the Genetic Code. *Mol. Cell. Biol.* **2002**, *22*, 3565–3576.

4. Reeves, M.A.; Hoffmann, P.R. The human selenoproteome: Recent insights into functions and regulation. *Cell. Mol. Life Sci.* **2009**, *66*, 2457–2478.

5. Bellinger, F.P.; Raman, A.V.; Reeves, M.A.; Berry, M.J. Regulation and function of selenoproteins in human disease. *Biochem. J.* **2009**, *422*, 11–22.

6. Burk, R.F.; Hill, K.E.; Motley, A.K. Selenoprotein metabolism and function: Evidence for more than one function for selenoprotein P. *J. Nutr.* **2003**, *133*, 1517s–1520s.

7. Duntas, L.H. The role of selenium in thyroid autoimmunity and cancer. *Thyroid* **2006**, *16*, 455–460.

8. Cardoso, B.R.; Roberts, B.R.; Bush, A.I.; Hare, D.J. Selenium, selenoproteins and neurodegenerative diseases. *Metallomics* **2015**, *7*, 1213–1228.

9. Moghadaszadeh, B.; Beggs, A.H. Selenoproteins and their impact on human health through diverse physiological pathways. *Physiology* **2006**, *21*, 307–315.

10. Köhrle, J.; Jakob, F.; Contempré, B.; Dumont, J.E. Selenium, the thyroid, and the endocrine system. *Endocr. Rev.* **2005**, *26*, 944–984.

11. Selinus, O.; Fordyce, F. Selenium deficiency and toxicity in the environment. In *Essentials of Medical Geology*; Springer Netherlands: Heidelberg, Germany, 2013; pp. 375–416.

12. Oldfield, J.E. *Selenium World Atlas: Updated Edition*; Selenium-Tellurium Development Association: Grimbergen, Belgium, 2002.

13. WHO. *Guidelines on Food Fortification with Micronutrients*; World Health Organization: Geneva, Switzerland, 2006.

14. Reilly, C. *Selenium in Food and Health*; Springer Science Media: New York, NY, USA, 2006.

15. Mangiapane, E.; Pessione, A.; Pessione, E. Selenium and selenoproteins: An overview on different biological systems. *Curr. Protein Pept. Sci.* **2014**, *15*, 598–607.

16. Rahmanto, A.S.; Davies, M.J. Selenium-containing amino acids as direct and indirect antioxidants. *IUBMB Life* **2012**, *64*, 863–871.

17. Patching, S.G.; Gardiner, P.H. Recent developments in selenium metabolism and chemical speciation: A review. *J. Trace Elem. Med. Biol.* **1999**, *13*, 193–214.

18. Steinbrenner, H.; Sies, H. Selenium homeostasis and antioxidant selenoproteins in brain: Implications for disorders in the central nervous system. *Arch. Biochem. Biophys.* **2013**, *536*, 152–157.

19. Hill, K.E.; Wu, S.; Motley, A.K.; Stevenson, T.D.; Winfrey, V.P.; Capecchi, M.R.; Atkins, J.F.; Burk, R.F. Production of selenoprotein P (Sepp1) by hepatocytes is central to selenium homeostasis. *J. Biol. Chem.* **2012**, *287*, 40414–40424.

20. Labunskyy, V.M.; Hatfield, D.L.; Gladyshev, V.N. Selenoproteins: Molecular pathways and physiological roles. *Physiol. Rev.* **2014**, *94*, 739–777.

21. Wastney, M.E.; Combs Jr, G.F.; Canfield, W.K.; Taylor, P.R.; Patterson, K.Y.; Hill, A.D.; Moler, J.E.; Patterson, B.H. A human model of selenium that integrates metabolism from selenite and selenomethionine. *J. Nutr.* **2011**, *141*, 708–717.

22. Schrauzer, G.N. Nutritional selenium supplements: Product types, quality, and safety. *J. Am. Coll. Nutr.* **2001**, *20*, 1–4.

23. Bierla, K.; Dernovics, M.; Vacchina, V.; Szpunar, J.; Bertin, G.; Lobinski, R. Determination of selenocysteine and selenomethionine in edible animal tissues by 2D size-exclusion reversed-phase HPLC-ICP MS following carbamidomethylation and proteolytic extraction. *Anal. Bioanal. Chem.* **2008**, *390*, 1789–1798.

24. Juniper, D.T.; Phipps, R.H.; Bertin, G. Effect of dietary supplementation with selenium-enriched yeast or sodium selenite on selenium tissue distribution and meat quality in commercial-line turkeys. *Animal* **2011**, *11*, 1751–1760.

25. Santhosh Kumar, B.; Priyadarsini, K.I. Selenium nutrition: How important is it? *Biomed. Prev. Nutr.* **2014**, *4*, 333–341.

26. Spears, J.W. Trace Mineral Bioavailability in Ruminants. *J. Nutr.* **2003**, *133*, 1506S–1509S.

27. Weiss, W.P. Selenium sources for dairy cattle. In Proceedings of Tri-State Dairy Nutrition Conference, Fort Wayne, IN, USA, May 2005; pp. 61–72.

28. Behne, D.; Kyriakopoulos, A. Mammalian selenium-containing proteins. *Ann. Rev. Nutr.***2001**, *21*, 453–473.

29. López-Alonso, M.R. Trace Minerals and Livestock: Not Too Much Not Too Little. *ISRN Vet. Sci.* **2012**.

30. Combs, G.F. Biomarkers of Selenium Status. *Nutrients* **2015**, *7*, 2209–2236.

31. Rayman, M.P. The use of high-selenium yeast to raise selenium status: how does it measure up? *Br. J. Nutr.* **2004**, *92*, 557–573.

32. Turanov, A.A.; Xu, X.-M.; Carlson, B.A.; Yoo, M.-H.; Gladyshev, V.N.; Hatfield, D.L. Biosynthesis of Selenocysteine, the 21st Amino Acid in the Genetic Code, and a Novel Pathway for Cysteine Biosynthesis. *Adv. Nutr.: An Int. Rev. J.* **2011**, *2*, 122–128.

33. Ma, S.; Caprioli, R.M.; Hill, K.E.; Burk, R.F. Loss of selenium from selenoproteins: Conversion of selenocysteine to dehydroalanine *in vitro*. *J. Am. Soc. Mass. Spectrom.* **2003**, *14*, 593–600.

34. Combs, G.F.; Scott Jr., M.L. Dietary requirements for vitamin E and selenium measured at the cellular level in the chick. *J. Nutr.* **1974,** *104*, 1292–1296.

35. Scott, M.L.; Noguchi, T.; Combs, G.F., Jr. New evidence concerning mechanisms of action of vitamin E and selenium. *Vitam. Horm.* **1974**, *32*, 429–444.

36. Diplock, A.T. The nutritional and metabollic roles of selenium and vitamin E. *Proc. Nutr. Soc.***1974**, *33*, 315–322.

37. National Research Council (US) Subcommittee on Selenium. *Selenium in Nutrition: Revised Edition*; National Academies Press: Washington, DC, USA, 1983.

38. Kinoshita, H.; Sugai, K.; Goto, Y.; Nonaka, I. Early onset distal muscular dystrophy. *Brain Dev.* **1995**, *17*, 206–209.

39. Sharp, B.A.; Van Dreumel, A.A.; Young, L.G. Vitamin E, selenium and methionine supplementation of dystrophogenic diets for pigs. *Can. J. Comp. Med.* **1972**, *36*, 398–402.

40. Sharp, B.A.; Young, L.G.; Van Dreumel, A.A. Dietary induction of mulberry heart disease and hepatosis dietetica in pigs. I. Nutritional aspects. *Can. J. Comp. Med.* **1972**, *36*, 371–376.

41. Jovanović, I.B.; Veličković, M.; Vuković, D.; Milanović, S.; Valčić, O.; Gvozdić, D. Effects of Different Amounts of Supplemental Selenium and Vitamin E on the Incidence of Retained Placenta, Selenium, Malondialdehyde, and Thyronines Status in Cows Treated with Prostaglandin F2α; for the Induction of Parturition. *J. Vet. Med.* **2013**.

42. Cheeke, P.R.; Dierenfeld, E.S. *Comparative animal nutrition and metabolism*; CABI: Wallingford, UK, 2010.

43. Finley, J.W. Bioavailability of selenium from foods. *Nutr. Rev.* **2006,** *64*, 146–151.

44. Fisinin, V.I.; Papazyan, T.T.; Surai, P.F. Producing selenium-enriched eggs and meat to improve the selenium status of the general population. *Crit. Rev. Biotechnol.* **2009**, *29*, 18–28.

45. Schrauzer, G.N.; Surai, P.F. Selenium in human and animal nutrition: resolved and unresolved issues. A partly historical treatise in commemoration of the fiftieth anniversary of the discovery of the biological essentiality of selenium, dedicated to the memory of Klaus Schwarz (1914–1978) on the occasion of the thirtieth anniversary of his death. *Crit. Rev. Biotechnol.* **2009**, *29*, 2–9.

46. Kieliszek, M.; Błażejak, S.; Gientka, I.; Bzducha-Wróbel, A. Accumulation and metabolism of selenium by yeast cells. *Appl. Microbiol. Biotechnol.* **2015**, *99*, 5373–5382.

47. Suhajda, A.; Hegoczki, J.; Janzso, B.; Pais, I.; Vereczkey, G. Preparation of selenium yeasts I. Preparation of selenium-enriched *Saccharomyces cerevisiae*. *J. Trace Elem. Med. Biol.* **2000**, *14*, 43–47.

48. Ouerdane, L.; and Mester, Z. Production and characterization of fully selenomethionine-labeled *Saccharomyces cerevisiae*. *J. Agric. Food Chem.* **2008**, *56*, 11792–11799.

49. Cantor, A.H.; Langevin, M.L.; Noguchi, T.; Scott, M.L. Efficacy of selenium in selenium compounds and feedstuffs for prevention of pancreatic fibrosis in chicks. *J. Nutr.* **1975**, *105*, 106–111.

50. Payne, R.L.; Southern, L.L. Comparison of inorganic and organic selenium sources for broilers. *Poult. Sci.* **2005**, *84*, 898–902.

51. da Silva, ICM; Ribeiro, A.M.L.; Canal, C.W.; Trevizan, L.; Macagnan, M.; Gonçalves, T.A.; Hlavac, N.R.C.; de Almeida, L.L.; Pereira, R.A. The impact of organic and inorganic selenium on the immune system of growing broilers submitted to immune stimulation and heat stress. *Rev. Bras. Cienc. Avic.* **2010**, *12*, 247–254.

52. Oliveira, T.F.B.; Rivera, D.F.R.; Mesquita, F.R.; Braga, H.; Ramos, E.M.; Bertechini, A.G. Effect of different sources and levels of selenium on performance, meat quality, and tissue characteristics of broilers. *J. Appl. Poult. Res.* **2014**, *23*, 15–22.

53. Tsuji, Y.; Mikami, T.; Anan, Y.; Ogra, Y. Comparison of selenohomolanthionine and selenomethionine in terms of selenium

distribution and toxicity in rats by bolus administration. *Metallomics* **2010**, *2*, 412–418.

54. Celi, P.; Selle, P.H.; Cowieson, A.J. Effects of organic selenium supplementation on growth performance, nutrient utilisation, oxidative stress and selenium tissue concentrations in broiler chickens. *Anim. Prod. Sci.* **2014**, *54*, 966–971.

55. Suzuki, K.T.; Doi, C.; Suzuki, N. Metabolism of 76Se-methylselenocysteine compared with that of 77Se-selenomethionine and 82Se-selenite. *Toxicol. Appl. Pharmacol.* **2006**, *217*, 185–95.

56. Anan, Y.; Ogra, Y. Toxicological and pharmacological a

57. Kaufman, N.; Klavins, J.V.; Kinney, T.D. Pancreatic damage induced by excess methionine.*Arch. Pathol.* **1960**, *70*, 331–337.

58. Henman, D.J.; (Riverlea Pty Ltd., Corowa, Australia). Personal communication, 2014.

# CHAPTER 8

## The Occurrence and Toxicity of Indospicine to Grazing Animals

**Mary T. Fletcher [1,*], Rafat A. M. Al Jassim [2] and A. Judith Cawdell-Smith [2]**

[1]Queensland Alliance for Agriculture and Food Innovation (QAAFI), The University of Queensland, Health and Food Sciences Precinct, Coopers Plains 4108, Australia

[2]School of Agriculture and Food Sciences, Faculty of Science, The University of Queensland, Gatton 4343, Australia

## ABSTRACT

Indospicine is a non-proteinogenic amino acid which occurs in *Indigofera* species with widespread prevalence in grazing pastures across tropical Africa, Asia, Australia, and the Americas. It accumulates in the tissues of grazing livestock after ingestion of *Indigofera*. It is a competitive inhibitor of arginase and causes both liver degeneration and abortion. Indospicine hepatoxicity occurs universally across animal species but the degree varies considerably between species, with dogs being particularly sensitive. The magnitude of canine sensitivity is such that ingestion of naturally indospicine-contaminated horse and camel meat has caused secondary poisoning of dogs, raising significant industry concern. Indospicine impacts on the health and production of grazing animals *per se* has been less widely documented. Livestock grazing *Indigofera* have a chronic and cumulative exposure to this toxin, with such exposure experimentally shown to induce both hepatotoxicity and embryo-lethal effects in cattle and sheep. In extensive pasture systems, where animals are not closely monitored, the resultant toxicosis may well occur after prolonged exposure but either be undetected, or even if detected not be attributable to a particular cause. Indospicine should be considered as a possible cause of animal poor performance, particularly reduced weight gain or reproductive losses, in pastures where *Indigofera* are prevalent.

## KEYWORDS

indospicine; indigofera; non-protein; hepatotoxic

## 1. INTRODUCTION

Indospicine (L-6-amidino-2-amino-hexanoic acid) is a non-proteinogenic amino acid found in*Indigofera* plant spp. which in recent decades has gained notoriety for its ability to form persistent and sometimes toxic residues in tissues of grazing animals. Ingestion of indospicine-contaminated horse meat [1] and more recently camel meat [2] has resulted in the secondary poisoning of dogs which are particularly sensitive to indospicine hepatotoxicosis. The impact of this plant toxin on the health and production of grazing animals *per se* has been less widely documented and is the focus of this review.

As an arginine analogue, indospicine (Figure 1) has the potential to interfere with a range of mammalian arginine metabolic pathways, as arginine is a precursor for the synthesis not only of proteins but also of nitric oxide, urea, polyamines, proline, glutamate, creatine and agmatine. Indospicine has been shown experimentally to cause both liver degeneration and also teratogenic and embryo-lethal effects [3,4], but the mechanism of this toxicity has never been fully elucidated.

## 2. PLANT SOURCES OF INDOSPICINE

Indospicine has been found only in *Indigofera* plant species, and then primarily restricted to a small number of species within this genus. As a cumulative, non-acute toxin it is possible that this toxin may be as yet unidentified in additional members of this genus of some 700 species which are distributed geographically in tropical Africa, Asia, Australia and North and South America.*Indigofera* plant species, like other Leguminosae, are high in protein, and are considered agronomically very desirable due to their ability to tolerate drought, floods and salinity [5]. Many*Indigofera* species in Africa and Asia have been utilized as forage, green manure in coffee, tea and rubber plantations, and as cover crops. Several species (notably *Indigofera spicata*) were initially introduced as pasture legumes into mainland USA [6] and into Australia [7], before their potential toxicity was recognised. Members of this genus range from prostrate or ascending perennial herbs (such as *I. spicata*) to shrubby species. These perennial shrubby species are deep-rooted and can represent valuable forage plants, due to their palatability, insect resistance and ability to respond with growth to small rainfall events [8], however indospicine has also been determined within some such species [9].

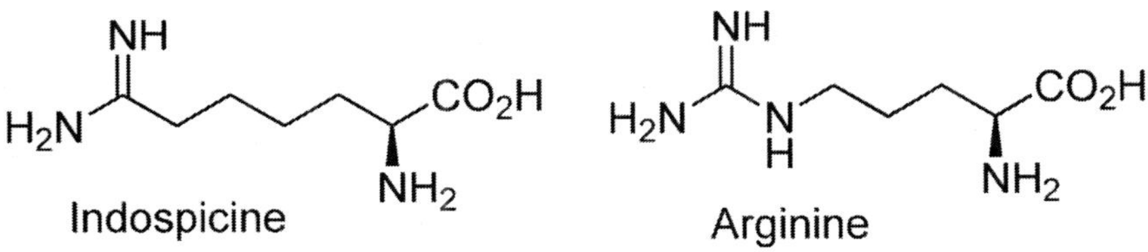


**Figure 1.** Indospicine an analogue of arginine.

## 2.1. Indospicine Content of Indigofera Species

This toxin was isolated as the first naturally occurring hepatotoxic amino acid in 1970 by Hegarty and Pound from *Indigofera spicata* hence the name indospicine [4]. Plant classification within this species has been confused over the years and the *I. spicata* complex now comprises both*I. spicata* (Forsk.) and *I. hendecaphylla* (Jacq.) (or *I. endecaphylla* auctt., orth. var.). These species names appear interchangeably in early literature and it was not until 1993 that these were shown to be 2 separate species [10]. It is thus difficult to ascertain with any certainty which of these species was actually used in early studies unless herbarium voucher specimens are examined [11]. Retrospective inspection of herbarium data has however established that the study by Hegarty and Pound utilized specimens of both species, and that indospicine was determined in both *I. spicata*and *I. hendecaphylla* in their work [11].

Since its discovery, indospicine content has been investigated in more than 30 different *Indigofera*species of both shrubby and herbaceous types with widely different results depending on both species and locality. Species in which high levels of indospicine have been determined in foliage (in excess of 500 mg/kg DM) include *I. hendecaphylla* [4,11], *I. linnaei* [12,13], *I. lespedezioides*[7,14,15], *I. spicata* [4,11,13,14,15,16], *I. vicioides* [9] and *I. volkensii* [7]. Indeed levels up to an exceptional 1.2% (DM basis) have been reported in *I. spicata* [7]. Species with low to moderate levels of indospicine (typically less than 100 mg/kg DM and sometimes up to some 300 mg/kg DM) include *I. amorphoides* [9], *I. arrecta* [9], *I. brevicalyx* [9], *I. circinella* [15], *I. coerulea* [9], *I. colutea* [13], *I. costata* [9], *I. cryptantha* [9], *I. heteroticha* [15], *I. hirsuta* [15], *I. linifolia* [13], *I. nigritana* [7] and *I. trita*[15].

It is perhaps worth noting that in a large proportion of the *Indigofera* species that have been analysed, indospicine has not been detected [7,13,15,17]. *Indigofera* species in this non-indospicine containing category include *I. adesmiifolia* [13], *I. alternans* [7], *I. antunesiana* [7], *I. astragalina* [7], *I. georgei* [13], *I. glandulosa* [7], *I. leucotricha* [13], *I. mucronata* [7], *I. oblongifolia* [13], *I. parviflora* [7,13],*I. praticola* [14], *I. schimperi* [7], *I. suffruticosa* [7], *I. tinctoria* [7,14], and *I. vohemarensis* [7]. Likewise Miller and Smith [17] measured indospicine in the defatted seed meals of "*I. endecaphylla*" (*I. spicata/I. hendecaphylla*) but did not detect indospicine in seed meal of 16 other *Indigofera* species namely *I. arrecta*, *I. brachystachya*, *I. densiflora*, *I. frutescens*, *I. hirsuta*, *I. jaliscensis*, *I. kirilowii*, *I. linifolia*, *I. oblongifolia*, *I. pilosa*, *I. schimperi*, *I. sphaerocarpa*, *I. suffruticosa*, *I. sulcata*, *I. tinctoria*, and *I. zoilingeriana*.

Wide variations in indospicine content have been reported within individual *Indigofera* species with some suggestion that levels are influenced by the locale in which the taxa is growing, and particularly wide ranges of concentration (more than 10 fold variations between individual samples) have been reported for plant samples of *I. linnaei* [13] and *I. lespedezioides* [14]. Growth stage also effects indospicine levels [4], with highest levels of indospicine found in harvested seed of *I. spicata* [4] and *I. linnaei* [12] and lesser concentrations found in other parts of the plant including tips, leaves and the stem.

In other examples of intra-species variation, indospicine was detected in *I. hirsuta* and *I. trita* by Charlwood [15] but not by other researchers

[7,13], and in *I. arrecta* and *I. amorphoides* by Hassen [9] but not by Aylward [7]. Aylward reported the "universality" of indospicine in *I. spicata* [7], whereas Charlwood [15] reported the absence of indospicine in 75% of seed samples of this species, and most remarkably detected only trace indospicine levels in the same seed accession [CPI 16110] that had contained 1240 mg/kg fresh weight when analysed by Hegarty and Pound [4]. The seed analysed by Hegarty and Pound in 1970 was green seed from fresh field-grown plants [4], while that analysed by Charlwood in 1984 was from a seed collection "collected from sites throughout Australia" and of unreported age [15]. This difference in seed source may well be a major contributor to the difference in reported results. Unsuccessful attempts were made to select for low-indospicine varieties of species with high agronomic potential such as *I. spicata*, and populations of this species were treated with a chemical mutagen in non-entirely successful attempts to develop indospicine-free cultivars of *I. spicata* [18].

## 2.2. Prevalence of Indospicine-Containing Indigofera Species

The 6 species with highest indospicine content *I. hendecaphylla*, *I. linnaei*, *I. lespedezioides*, *I. spicata*,*I. vicioides* and *I. volkensii*, are all prostrate or semi-prostrate herbs that occur in pasture and are potentially grazed by livestock. *I. hendecaphylla*, *I. linnaei*, *I. lespedezioides*, *I. spicata*, and *I. volkensii*are relatively closely phylogenetically aligned within the Tethyan Clade, whereas *I. vicioides* is distinct within the Pantropical clade [19].

*I. spicata* and *I. hendecaphylla* are widely spread across tropical regions. *I. hendecaphylla* has a native range in Africa, Comoros, Madagascar, Reunion, Asia to Papua New Guinea and the Philippines, but it has been introduced and is now naturalized extensively in Florida (USA), French Polynesia and also Australia [20]. Similarly, *I. spicata* has a native range in Africa, Madagascar, Mauritius and Yemen, and has been naturalized extensively in Australia, Hawaiian islands, Japan, New Caledonia, Micronesia, Cook Islands among other places [21]. *I. hendecaphylla*and *I spicata* are often confused in literature, and it was only recently established that the species of*Indigofera* found in Hawaii was *I. spicata* [22]. Even though their toxicity has been recognized for more than 70 years [6], these species are still utilized as forage in their native range [23]. *I. linnaei*has a less extensive global distribution and is native across northern Australia, India, Indochina through Malesia, Melanesia to New Guinea [24]. This species has a seasonally variable prevalence in pasture, and can at times be a predominant plant in some arid regions of central and northern Australia. With a strong tap root the perennial *I. linnaei* is able to withstand dry conditions and responds quickly after rain [25]. It is most abundant at the start of the summer wet season when first rains stimulate the growth of this plant but are not sufficient to cause the germination of annual grasses and herbage. Horses in this region suffer from the neurological syndrome Birdsville horse disease following consumption of this plant [25], but cattle and sheep are seemingly unaffected by *I. linnaei* under normal grazing circumstances [26]. *I. lespedezioides* is restricted to the Americas and is native from Mexico and throughout South America [27] where the plant, particularly the root, is used medicinally as a febrifuge and stomachic (Venezuela); to halt diarrhea (Honduras); and as a treatment for gonorrhea, leucorrhea, jaundice, and epilepsy (Brazil) [28]. In Brazil, the roots and leaves are employed in the Mato Grosso to poison fish [28]. This species is responsible for a neurological syndrome in horses in Brazil [29] and is reported to be poisonous to livestock in Bolivia [30] and cattle in Brazil [29].

*I. vicioides* and *I. volkensii* have less extensive distributions and there are no literature reports of toxicity in livestock consuming these two species. *I. vicioides* is a spreading annual or perennial herb with native range in Africa and India [31] and *I. volkensii* has a more limited native range in east and north east Africa [32].

## 3. ETIOLOGY OF INDOSPICINE TOXICOSIS

Indospicine toxicosis manifests itself in two ways, acting as both a hepatotoxin with associated fat accumulation in liver and secondly as a teratogen and abortifacient. These effects were first observed in livestock consuming *Indigofera* plant material, with indospicine subsequently identified as the responsible toxin through *in vivo* rodent toxicity studies [3,4].

*I. spicata* plant extracts were shown to induce a site-specific cleft of the secondary palate, generalized somatic dwarfism, and intrauterine embryolethality [33,34]. The abortive effect was also reported in pregnant rabbit does where feeding as little as 5% *I. spicata* in daily rations resulted in 100% stillborn offspring when fed in the last 15 days of pregnancy [35]. Pure indospicine administered orally at day 13 of gestation in pregnant rats caused cleft palate in more than 80% of viable fetuses exposed with histological features identical to those produced by the crude *I. spicata* extract, and also high embryolethality [3]. No other physical defects were observed in the embryopathic syndrome. Rat pups born with cleft palate induced by *Indigofera* did not live for more than a few hours after birth, with this observation attributed to either inhalation of milk into the lungs or a secondary metabolic toxic effect [33].

Indospicine also interferes with hepatic protein metabolism. Subcutaneous injection of indospicine into mice produced fat accumulation and cytological changes in the liver, which was inhibited by simultaneous injection of arginine [4]. In *in vivo* rat studies, oral dosing caused a pronounced inhibition of incorporation of $^{14}$C-leucine into hepatic and serum protein, with this inhibition of protein synthesis then followed by a marked rise in the level of liver triglycerides [36]. Incorporation of $^{14}$C-arginine into protein was depressed in cell-free fractions from the livers of rats treated with indospicine and also on addition of indospicine *in vitro* [37]. Increasing the level of arginine in the mixture decreased the magnitude of this effect. It was shown also that the formation of $^{14}$C-arginyl-tRNA was inhibited by indospicine, but not the formation of $^{14}$C-leucinyl-tRNA. Indospicine was concluded to be a competitive antagonist of arginine preventing arginine being incorporated into tRNA as a prelude to ribosomal polypeptide synthesis. Indospicine was also shown to be a competitive inhibitor of rat liver homogenate arginase with the affinity of arginase for indospicine being similar to that of the enzyme for arginine [38]. Inhibition of arginase negatively impacts aminoacylation of arginine and adversely leads to dysfunction of urea cycle. It was considered that other enzymes involved in arginine metabolism may be similarly affected by indospicine. The effect of indospicine on multiplying cells was tested by measuring the incorporation of $^{3}$H-thymidine into newly-formed DNA using cultures of human lymphocytes stimulated with phytohaemagglutinin. Indospicine inhibited DNA synthesis particularly at early stages of incubation, and the inhibition was reversible by arginine proportional to the relative concentrations of the two substances [39]. This inhibition was attributed to blocking of protein synthesis which in turn is essential for DNA synthesis. Inhibition of DNA synthesis was also demonstrated by addition of indospicine to mouse bone marrow cell cultures, and this inhibition could also be prevented by supplementation by arginine [40].

In addition to indospicine, some *Indigofera* species contain a further non-proteinogenic amino acid, canavanine, also related to arginine. Canavanine can function in all enzymatic reactions for which arginine is a substrate and does not induce the same toxicosis as seen with indospicine [4]. Additionally some *Indigofera* species contain 3-nitropropanoic acid (3-NPA), present in the plant material as glucose esters. Independent rat trials demonstrated that 3-NPA is a neurotoxin and does not cause the hepatotoxic effects linked with indospicine [41]. Horses (but not ruminants) develop a neurological syndrome thought to be related to 3-NPA effects as discussed below.

## 4. PERSISTENCE OF INDOSPICINE IN ANIMAL TISSUES

Indospicine is an unusual amino acid in that it is not incorporated into protein but exists in both plant and animal tissues as the free amino acid [42]. It is seemingly not readily metabolized by mammalian systems and accumulates as persistent residues in all tissues of animals consuming a diet containing the amino acid. Indospicine has been shown to accumulate in the tissues of horses [1,42,43], cattle [44] and goats [45] consuming *I. linnaei*, and in rabbits fed *I. spicata* seed [12]. The persistence of these residues has been demonstrated in experimental studies and tissue indospicine levels were still detectable several months after the cessation of feeding *Indigofera* plant material in horses [43], cattle [44], goats [45] and rabbits [12]. The cumulative and persistent nature of indospicine may well contribute to the long term toxic effects seen in field studies.

## 5. INDOSPICINE EFFECTS ON ANIMALS CONSUMING *INDIGOFERA* PLANTS

The toxicity of *Indigofera* plants to livestock and other animals has been recorded over many decades with differing manifestations of indospicine toxicosis depending on animal species.

## 5.1. Cattle

Cattle consuming *Indigofera* have been reported to experience both hepatotoxic effects with associated liver lesions, and also teratogenic and embryo-lethal effects [35]. Cattle can, however, seemingly tolerate relatively small proportions of *Indigofera* with no observable ill effects. In Australia, cattle are reported to graze *I. linnaei* with no ill effect under normal grazing circumstances [26]. In Kenya, researchers, in 1949, reported various *Indigofera* including "*I. endecaphylla*" (*I. spicata* or *I. hendecaphylla*) were palatable to cattle with no toxicity reported [46]. In Hawaii, studies in the 1940s showed no adverse effects on young cattle in 10 years of short-interval pasture trials with relatively small proportions of the *I. spicata* [35]. However as reported by Nordfeldt *et al.*, in 1952, when the concentration of the legume in the diet was increased and the trials were continued for longer periods of time, signs of toxicity began to appear. Cattle fed a diet containing increasing percentages from 30% to 100% fresh *I. spicata* for 15 days gradually became inappetent and lost weight. The central nervous system was reportedly affected in some animals that appeared sluggish and drowsy, walking in circles with low head carriage, although these signs usually disappeared within one or two days. Animals fed for longer time periods showed depressed weight gain and liver lesions [35]. Loss in bodyweight and liver lesions were also reported in calves fed the fresh "*I. endecaphylla*" as a third of their diet for 2 weeks in Ceylon (Sri Lanka) in 1953 [47]. These authors recorded elevated levels of albumin in the urine and swollen, congested livers and pale kidneys on necropsy. Histological examination of the livers revealed periacinar necrosis, haemorrhage, and distention of the capillaries and sinusoids.

In addition to weight loss and hepatoxicity, cattle consuming *Indigofera* have also been reported to suffer reproductive loses, with "*I. endecaphylla*" (*I. spicata* or *I. hendecaphylla*) recognised as causing abortion in cattle in Africa as early as 1937 [48]. In trials in Hawaii in 1949, Nordfeldt reported *I. spicata* caused abortion or stillbirths in pregnant cows and heifers fed *I. spicata* as 16% to 52% of the diet for up to 44 days [35,49]. Dead or stillborn calves were born prematurely by as much as 100 days after heifers were fed green *I. spicata* for 25 days as a pasture supplement. In Fiji, abortions and placental retention were reported in all 4 cows fed between 25% and 100% *I. spicata*for one to three months in 1959 [50]. No autopsy information was provided for dead or stillborn calves and it is not known whether the cleft palate seen in rats fed indospicine was present. We are unable to locate any primary literature that describes cleft palate in calves as a result of maternal consumption of *Indigofera* or indospicine, however the Merck

Veterinary Manual states that *I. spicata* is suspected of causing cleft palate in calves [51]. It has been further suggested that the widespread occurrence and utilization of *Indigofera* by livestock in Australian rangelands may also be an unattributed cause of malformations in domestic livestock [52]. *I. linnaei* has a widespread distribution in Australian rangelands [53], and given the extensive nature of livestock production systems in arid regions of central and northern Australia, pre-natal and peri-natal losses due to maternal consumption of indospicine may well occur and contribute to observed reproduction losses in these regions without any specific attribution. Causes of a large proportion of both pre- and peri-natal losses in these extensive herds are unknown [54]. In a recent study of factors affecting calf mortality in this region, 67% of calves that died did so within a week of their birth, with cause of death most frequently recorded as unknown [55].

## 5.2. Sheep

Fewer *Indigofera* feeding experiments have been conducted with sheep than cattle, but it would seem that sheep also experience both the hepatotoxic effects of indospicine with associated liver lesions, and embryo-lethality. The clinical signs of toxicoses in sheep fed *I. spicata* are largely the same as those seen in cattle [35,56]. Nordfeldt reported a more pronounced toxic effect on sheep than on cattle in feeding trials in Hawaii in 1952. In addition to inappetence, weight loss and drowsiness, sheep also exhibited an ocular discharge, a discharge from the nostrils and, in later stages, corneal opacity. Anorexia developed in sheep after two weeks on an *I. spicata* diet with deaths occurring after 28 days. Most of the sheep exhibited general weakness and an unsteady gait, while one animal developed severe diarrhea and one pregnant ewe aborted its foetus. Post-mortem examination revealed fatty degeneration of the liver. In later Australian trials, sheep fed *I. spicata ad lib* were reported to suffer inappetence, weight loss and drowsiness, with liver lesions seen post-mortem [56].

Like cattle, sheep can however, seemingly tolerate relatively small proportions of *Indigofera* with no observable ill effects. In Australia, sheep are reported to graze *I. linnaei* with no apparent ill effect [26]. In Kenya researchers in 1949 reported that various *Indigofera* spp. including "*I. endecaphylla*" (*I. spicata* or *I. hendecaphylla*) were palatable to sheep with no toxicity reported [46]. Feeding trials in India suggested *I. cordifolia* and *I. linnaei* (then called *I. enneaphylla*) were highly palatable feeds for sheep [57], and these plants are still considered valuable pasture constituents in India [58].

In contrast, sheep fed unrestricted amounts of *I. teysmannii* (*I. zoilingeriana*) leaves for a period of 24 days in India suffered hepatotoxic and nephrotoxic changes [59]. Rams were jaundiced with significant serum concentrations of albumin, glucose, bile salts and bilirubin. Fat deposits were gelatinized in livers at necropsy and histological examinations revealed fatty changes in the liver and mononuclear infiltrates with areas of necrosis and fibrosis, and regenerative foci. Lesions were also present in the gastrointestinal tract, heart, spleen, lungs and kidney. However the indospicine content of this species is not known, indeed Miller reported the absence of indospicine in seeds of this species [17]. It is thus not clear whether the reported toxicity of *I. teysmannii* can be attributed to indospicine.

## 5.3. Horses

Horses consuming *I. linnaei* and *I. spicata* appear resistant to the hepatotoxic effect of indospicine [1,11]. Microscopic examination reveals only mild liver lesions consistent with descriptions of indospicine-related hepatopathy, including periacinar and periportal lymphocytic infiltrations, hydropic degeneration of mid-zonal hepatocytes and multifocal periacinar necrosis [11]. Horses however don't escape embryolethality indospicine effects, and abortion is commonly observed in mares consuming *I. lespedeziodes* in Brazil [29].

Equine neurological syndromes known as "Birdsville horse disease" in Australia [11] and "Grove disease" in USA [60] are attributed to consumption of *Indigofera* plant species but do not relate to indospicine toxicosis. *Indigofera* species particularly *I. hendecaphylla*, *I. linnaei* and *I. spicata*have been shown to contain glucose esters of 3-nitropropanoic acid (3-NPA) co-occurring with indospicine, and 3-NPA is believed to be the toxin responsible for the neurological syndrome seen in horses (but not ruminants) consuming these plants [11,61,62,63]. In non-ruminants, esters of 3-NPA are hydrolysed by mammalian esterase activity in the upper gastrointestinal tract but rapid absorption of the nitroacid precludes its metabolism by microbes in the hindgut [64]. In ruminants 3-NPA can be metabolised and detoxified by ruminal bacteria before absorption from the intestine, and hence the lesser toxicity of 3-NPA in these species. *I. lespedeziodes* also contains both indospicine and 3-NPA [14,29], but it has been queried whether detected levels of 3-NPA (up to 2.5 mg/g) were sufficient to cause the neurological syndrome seen in Brazil and these authors suggested that indospicine could be responsible for the neurological syndrome [29]. Notably however none of the tested plant samples were directly related to poisoning incidents.

There is no evidence that indospicine causes these neurological syndromes in horses, but we propose that is possible that indospicine acting as an arginine antagonist could attenuate the 3-NPA effects in horses. Creatine has been demonstrated to offer a protection against the neurological effects of 3-NPA [65], and it is thus conceivable that an indospicine-induced arginine (and resultant creatine) deficiency could exacerbate the neurological effects of 3-NPA.

Interestingly, even though unaffected by the hepatotoxic effects of indospicine, horses consuming *Indigofera* plants are known to accumulate indospicine in their tissues [1,11]. Indospicine accumulation occurs in all tissues including muscle and is residual. Fatal hepatotoxicity has occurred in dogs fed indospicine-contaminated meat from horses that had grazed *I. linnaei* in central Australia [1]. As a consequence, the Northern Territory of Australia *Meat Industries Act* [66] bans the slaughter of horses, donkeys, mules or hinnies for human consumption if the animals are exhibiting signs of Birdsville horse disease (*I. linnaei* poisoning), or for pet food if these animals originate from an area in which Birdsville disease occurs.

## 5.4. Camels

*I. linnaei* is reportedly a preferred food plant for feral camels present in arid Australian rangelands [67]. There are no field reports of impacts on camel health, but canine deaths have resulted from the consumption of indospicine-contaminated camel meat [2]. Camels fed *I. spicata*for 35 days accumulated indospicine in their tissues but showed no clinical signs [68]. At autopsy only very mild liver degeneration was observed.

## 5.5. Goats

*I. hochstetteri* has been demonstrated to be toxic to goats in Sudan, with animals developing inappetence, weight loss, anaemia and hind limb weakness followed by incoordination, recumbency and death [69]. Fat accumulation and congestion of the liver were reported at necropsy as well as damage to other tissues, especially the kidneys, intestine and central nervous system. Histological lesions included fatty degeneration of the liver. This is suggestive of indospicine toxicity, even though indospicine content of the plant had not been established.

An Angora goat fed dried *I. linnaei* at an average dose rate of 0.2 mg indospicine/kg BW/day for 52 days accumulated indospicine in plasma and liver but did not develop liver lesions [45]. A second goat was fed at dose rates of 0.2 mg to 0.6 mg indospicine/kg BW/day over a 30 week period and demonstrated the persistence of plasma indospicine levels with a half-life of some 2 weeks after cessation of *Indigofera* consumption.

## 5.6. Poultry/Birds

Following the reports of *I. spicata* toxicity in grazing livestock by Nordfeldt [35], feeding trials with 11 different *Indigofera* species were undertaken with poultry chicks [70]. Of the species tested only one induced toxicity. Consumption of 5% "*I. endecaphylla*" (*I. spicata* or *I. hendecaphylla*) meal from various global sources resulted in decreased growth rate and paralysis of the neck, legs and wings, commonly followed by death [70]. Subsequent investigation established that the toxin responsible for producing these neurological signs in chicks was 3-NPA and that the severity of the toxic signs was related to the 3-NPA content of the plant material [63,71,72,73]. Chickens were intoxicated by pure 3-NPA and by the leaf and stem of *I. spicata* containing 3-NPA as well as indospicine, but were unaffected by *I. spicata* seed, which contains indospicine, but no 3-NPA [74]. This evidence suggests that chickens are not susceptible to the hepatotoxic effects of indospicine.

# 6. CONCLUSIONS

Indospicine is not an acute toxin, but it is present in a number of *Indigofera* species which have a somewhat ubiquitous presence in grazing pastures across tropical regions from Africa, Asia, Australia, the Americas and islands of both Pacific and Indian Oceans. Livestock can thus have a chronic and cumulative exposure to this toxin, with such exposure experimentally shown to induce both hepatotoxicity and embryonic losses in cattle and sheep. In extensive pasture systems, where animals are not closely and individually monitored, the resultant toxicosis may well occur after prolonged exposure but either not be detected/identified or even if detected then be attributed to an unknown cause. The literature would suggest that for cattle or sheep grazing pastures where *Indigofera* (particularly *I. spicata*, *I. hendecaphylla*, and *I. linnaei*) are prevalent, indospicine should be considered as a possible cause of poor performance, particularly where reduced weight gain or reproductive losses are involved.

# AUTHOR CONTRIBUTIONS

Mary T. Fletcher reviewed the literature and wrote the initial draft of the manuscript with assistance from A. Judith Cawdell-Smith and Rafat Al Jassim. All three authors reviewed and edited the final submission.

# CONFLICTS OF INTEREST

The authors declare no conflict of interest.

# REFERENCES

1. Hegarty, M.P.; Kelly, W.R.; McEwan, D.; Williams, O.J.; Cameron, R. Hepatotoxicity to dogs of horse meat contaminated with indospicine. *Aust. Vet. J.* **1988**, *65*, 337–340.

2. FitzGerald, L.M.; Fletcher, M.T.; Paul, A.E.H.; Mansfield, C.S.; O'Hara, A.J. Hepatotoxicosis in dogs consuming a diet of camel meat contaminated with indospicine. *Aust. Vet. J.* **2011**, *89*, 95–100.

3. Pearn, J.H.; Hegarty, M.P. Indospicine—The teratogenic factor from *Indigofera spicata* extract causing cleft palate. *Br. J. Exp. Pathol.* **1970**, *51*, 34–36.

4. Hegarty, M.P.; Pound, A.W. Indospicine, a hepatotoxic amino acid from *Indigofera spicata*: Isolation, structure, and biological studies. *Aust. J. Biol. Sci.* **1970**, *23*, 831–842.

5. Skerman, P.J.; Cameron, D.G.; Riveros, F. *Tropical Forage Legumes*, 2nd ed.; Food and Agricultural Organisation of the United Nations: Rome, Italy, 1988.

6. Emmel, M.W.; Ritchey, G.E. The toxicity of *Indigofera endecaphylla* Jacq. for rabbits. *J. Am. Soc. Agron.* **1941**, *33*, 675–677.

7. Aylward, J.H.; Court, R.D.; Haydock, K.P.; Strickland, R.W.; Hegarty, M.P. *Indigofera* species with agronomic potential in the tropics. Rat toxicity studies. *Aust. J. Agric. Res.* **1987**, *38*, 177–186.

8. Hassen, A.; Rethman, N.F.G.; van Niekerk, W.A.; Tjelele, T.J. Influence of season/year and species on chemical composition and *in vitro* digestibility of five *Indigofera* accessions. *Anim. Feed Sci. Technol.* **2007**, *136*, 312–322.

9. Hassen, A.; Rethman, N.F.G.; Apostolides, Z.; van Niekerk, W.A. Forage production and potential nutritive value of 24 shrubby *Indigofera* accessions under field conditions in South Africa. *Trop. Grasslands* **2008**, *42*, 96–103.

10. Du Puy, D.J.; Labat, J.N.; Schrire, B.D. The separation of two previously confused species in the *Indigofera spicata* complex (Leguminosae: Papilionoideae). *Kew Bull.* **1993**, *48*, 727–733.

11. Ossedryver, S.M.; Baldwin, G.I.; Stone, B.M.; McKenzie, R.A.; van Eps, A.W.; Murray, S.; Fletcher, M.T. *Indigofera spicata* (creeping indigo) poisoning of three ponies. *Aust. Vet. J.* **2013**,*91*, 143–149.

12. Pollitt, S. Residue Implications of Indospicine, a Toxic, Non-Protein Amino Acid. Ph.D. Thesis, Department of Physiology and Pharmacology, The University of Queensland, Brisbane, Australia, 2001.

13. Tan, E.T.T.; Silcock, R.G.; Al Jassim, R.; D'Arcy, B.R.; Fletcher, M.T.; The University of Queensland, Brisbane, Queensland, Australia. Unpublished work. 2015.

14. Gardner, D.; Riet-Correa, F. Analysis of the toxic amino acid indospicine by liquid chromatography-tandem mass spectrometry. *Int. J. Poisonous Plant Res.* **2011**, *1*, 20–27.

15. Charlwood, B.V.; Morris, G.S.; Grenham, M.J. A chemical database for the Leguminosae. In*Database in Systematics*; Alkin, R., Bisby, F.A., Eds.; Academic Press: London, UK, 1984; pp. 201–208.

16. Aylward, J.H. Screening legumes for toxicity. In *CSIRO Division of Tropcal Crops and Pastures Annual Report 1981–1982*; CSIRO: New South Wales, Australia, 1982; pp. 80–81.

17. Miller, R.W.; Smith, C.R., Jr. Seeds of *Indigofera* species. Their content of amino acids that may be deleterious. *J. Agric. Food Chem.* **1973**, *21*, 909–912.

18. Duke, J.A. *Handbook of Legumes of World Economic Importance*; Plenum Press: New York, NY, USA, 1981.

19. Schrire, B.D.; Lavin, M.; Barker, N.P.; Forest, F. Phylogeny of the tribe Indigofereae (Leguminosae–Papilionoideae): Geographically structured more in succulent-rich and temperate settings than in grass-rich environments. *Am. J. Bot.* **2009**, *96*, 816–852.

20. PIER (Pacific Islands Ecosystems at Risk). *Indigofera Hendecaphylla* Jacq., Fabaceae. Available online: http://www. hear. org/pier/ species/indigofera_hendecaphylla.htm (accessed on 9 July 2015).

21. PIER (Pacific Islands Ecosystems at Risk). *Indigofera spicata* Forssk., Fabaceae. Available online: http://www.hear.org/pier/ species/indigofera_spicata.htm (accessed on 9 July 2015).

22. Wagner, W.L.; Herbst, D.R.; Lorence, D.H. Flora of the Hawaiian Islands website. Available online: http://botany.si.edu/ pacificislandbiodiversity/ hawaiianflora/index.htm (accessed on 9 July 2015).

23. Bezabih, M.; Pellikaan, W.F.; Tolera, A.; Khan, N.A.; Hendriks, W.H. Chemical composition and *in vitro* total gas and methane production of forage species from the Mid Rift Valley grasslands of Ethiopia. *Grass Forage Sci.* **2014**, *69*, 635–643. [Google Scholar] [CrossRef]

24. GRIN (USDA Germplasm Resources Information Network). Taxon: *Indigofera linnaei* Ali. Available online: http://www.ars-grin.gov/cgi-bin/npgs/html/taxon.pl?70382 (accessed on 9 July 2015).

25. Gracie, A.; Brown, A.; Saville, P. AgNote: Birdsville Disease (Northern Territory Government). Available online: http://www.nt.gov.au/d/Content/File/p/Anim_Dis/633.pdf(accessed on 9 July 2015).

26. Dowling, R.M.; McKenzie, R.A. *Poisonous Plants: A Field Guide*; Department of Primary Industries: Brisbane, Australia, 1993.

27. GRIN (USDA Germplasm Resources Information Network). Taxon: *Indigofera lespedezioide*s Kunth. Available online: http://www.ars-grin.gov/cgi-bin/npgs/html/taxon.pl?20025(accessed on 9 July 2015).

28. Hastings, R.B. Medicinal legumes of Mexico: Fabaceae, Papilionoideae, Part One. *Econ. Bot.* **1990**, *44*, 336–348.

29. Lima, E.F.; Riet-Correa, F.; Gardner, D.R.; Barros, S.S.; Medeiros, R.M.T.; Soares, M.P.; Riet-Correa, G. Poisoning by *Indigofera lespedezioides* in horses. *Toxicon* **2012**, *60*, 324–328.

30. Altschul, S.V.R. *Drugs and Foods from Little-Known Plants: Notes in Harvard University Herbaria*; Harvard University Press: Cambridge, MA, USA, 1973.

31. GRIN (USDA Germplasm Resources Information Network). Taxon: *Indigofera vicioides* Jaub. & Spach. Available online: http://www.ars-grin.gov/cgi-bin/npgs/html/tax_search.pl(accessed on 9 July 2015).

32. GRIN (USDA Germplasm Resources Information Network). Taxon: *Indigofera volkensii* Taub. Available online: http://www.ars-grin.gov/cgi-bin/npgs/html/tax_search.pl (accessed on 9 July 2015).

33. Pearn, J.H. Report of a new site-specific cleft palate teratogen. *Nature* **1967**, *215*, 980–981.

34. Pearn, J.H. Studies on a site-specific cleft palate teratogen. The toxic extract from *Indigofera spicata* Forssk. *Br. J. Exp. Pathol.* **1967**, *48*, 620–626.

35. Nordfeldt, S.; Henke, L.A.; Morita, K.; Matsumoto, H.; Takahash, M.; Younge, O.R.; Willers, E.H.; Cross, R.F. *Feeding Tests with*

*Indigofera Endecaphylla Jacq. (Creeping Indigo) and Some Observations on its Poisonous Effects on Domestic Animals*; Technical Bulletin No. 15; Hawaii Agriculture Experiment Station, University of Hawaii: Honolulu, HI, USA, 1952; pp. 5–23.

36. Christie, G.S.; Madsen, N.P.; Hegarty, M.P. Acute biochemical changes in rat liver induced by the naturally-occurring amino acid indospicine. *Biochem. Pharmacol.* **1969**, *18*, 693–700.

37. Madsen, N.P.; Christie, G.S.; Hegarty, M.P. Effect of indospicine on incorporation of L-arginine-$^{14}C$ into protein and transfer ribonucleic acid by cell-free systems from rat liver.*Biochem. Pharmacol.* **1970**, *19*, 853–857.

38. Madsen, N.P.; Hegarty, M.P. Inhibition of rat liver homogenate arginase activity *in vitro* by the hepatotoxic amino acid indospicine. *Biochem. Pharmacol.* **1970**, *19*, 2391–2393.

39. Christie, G.S.; de Munk, F.G.; Madsen, N.P.; Hegarty, M.P. Effects of an arginine antagonist on stimulated human lymphocytes in culture. *Pathology* **1971**, *3*, 139–144.

40. De Munk, F.G.; Christie, G.S.; Hegarty, M.P. The effects of indospicine on bone marrow cells in liquid culture. *Pathology* **1972**, *4*, 133–137.

41. Hutton, E.M.; Windrum, G.M.; Kratzing, C.C. Studies on the toxicity of *Indigofera endecaphylla*: II. Toxicity for mice. *J. Nutr.* **1958**, *65*, 429–440.

42. Pollitt, S.; Hegarty, M.P.; Pass, M.A. Analysis of the amino acid indospicine in biological samples by high performance liquid chromatography. *Nat. Toxins* **1999**, *7*, 233–240.

43. Hegarty, M.P. Non-metallic chemical residues in toxic plants with potential importance to animal and human health. In *Vet Update 92*; Osborne, H.G., Ed.; Continuing Professional Education, University of Queensland: Brisbane, Australia, 1992; pp. 323–332.

44. Arid Zone Research Institute. Indospicine in beef. In *Northern Territory Department of Primary Industries and Fisheries Technical Annual Report 1987–1988*; Technical Bulletin No. 150; NTDPIF: Alice Springs, Australia, 1989; p. 39.

45. Young, M.P. Investigation of the Toxicity of Horsemeat due to Contamination by Indospicine. Ph.D. Thesis, School of Veterinary Science, University of Queensland, Brisbane, Australia, 1992.

46. Bogdan, A.V. Observations on palatability of some leguminous plants of Kenya. *East Afr. Agric. J.* **1949**, *15*, 38–40.

47. Jeganathan, P. Toxic effects of feeding *Indigofera endecaphylla* (Jacq.) to calves. *Ceylon Vet. J.***1953**, *1*, 83–85.

48. Dalziel, J.M. *The Useful Plants of West Tropical Africa*; The Crown Agents for the Colonies: London, UK, 1937; pp. 243–247.

49. Nordfeldt, S.; Young, O.R. Toxicity of creeping indigo to livestock. *Hawaii Agric. Expt. Station Univ. Hawaii Prog. Notes* **1949**, *55*, 1–2.

50. Yelf, J.D. The toxicity of creeping indigo in Fiji. *Agric. J.* **1959**, *29*, 9–10.

51. Maxwell, H.S.; Nettleton, P.; Kirkland, P.D. Overview of congenital and inherited anomalies.*Merck Veterinary Manual*, Available online: http://www.merckvetmanual.com/mvm/(accessed on 9 July 2015).

52. Keeler, R.F. Plant metabolites that are teratogenic in offspring and toxic in the dam. *Toxicon***1983**, *21*, 221–225.

53. Australia's Virtual Herbarium. Available online: http://avh. chah. org.au/ (accessed on 9 July 2015).

54. Burns, B.M.; Fordyce, G.; Holroyd, R.G. A review of factors that impact on the capacity of beef cattle females to conceive, maintain a pregnancy and wean a calf—Implications for reproductive efficiency in northern Australia. *Anim. Reprod. Sci.* **2010**, *122*, 1–22.

55. Bunter, K.L.; Johnston, D.J.; Wolcott, M.L.; Fordyce, G. Factors associated with calf mortality in tropically adapted beef breeds managed in extensive Australian production systems.*Anim. Prod. Sci.* **2014**, *54*, 25–36.

56. Maskasame, C. Toxicity and Nutritional Value of Some Promising Pasture Legumes in Rats and Sheep. Master's Thesis, Department of Physiology and Pharmacology, The University of Queensland, Brisbane, Australia, 1984.

57. Nath, K.; Malik, N.; Singh, O. Chemical composition and nutritive value of *Indigofera enneaphylla* and *I. cordifolia* as sheep feeds. *Aust. J. Exp. Agric.* **1971**, *11*, 178–185.

58. Srinivas, B.; Swain, N. Seasonal dynamics in vegetation and rumen microbial nitrogen production and nutritional status of grazing sheep in a semiarid rangeland in eastern Rajasthan, India. *Grassland Sci.* **2011**, *57*, 219–224.

59. Krishna, L.; Vaid, J.; Singh, B. Pathological study on *Indigofera teysmanni* toxicity in sheep.*Indian J. Comp. Microbiol. Immunol. Infect. Dis.* **1986**, *7*, 14–17.

60. Morton, J.F. Creeping indigo (*Indigofera spicata* Forsk.) (Fabaceae): A hazard to herbivores in Florida. *Econ. Bot.* **1989**, *43*, 314–327.

61. Majak, W.; Benn, M.; McEwan, D.; Pass, M.A. Three nitropropanoyl esters of glucose from*Indigofera linnaei*. *Phytochemistry* **1992**, *31*, 2393–2395.

62. Williams, M.C. Nitro compounds in *Indigofera* species. *Agron. J.* **1981**, *73*, 434–436.

63. Cooke, A.R. The toxic constituent of *Indigofera endecaphylla*. *Arch. Biochem. Biophys.* **1955**, *55*, 114–120.

64. Majak, W.; Pass, M.A. Aliphatic nitrocompounds. In *Toxicants of Plant Origin, Volume II, Glycosides*; Cheeke, P.R., Ed.; CRC Press: Boca Raton, FL, USA, 1989; pp. 143–159.

65. Shear, D.A.; Haik, K.L.; Dunbar, G.L. Creatine reduces 3-nitropropionic-acid-induced cognitive and motor abnormalities in rats. *NeuroReport* **2000**, *11*, 1833–1837.

66. Northern Territory Dept Primary Industries Fisheries & Mines. Northern Territory of Australia Meat Industries Act. Available online: http://www.austlii.edu.au/au/legis/nt/consol_act/ mia184.txt/ cgi-bin/download.cgi/download/au/legis/nt/consol_act/mia184.rtf# _Toc263691324 (accessed on 9 July 2015).

67. Dörges, B.; Heucke, J.; Dance, R. *The Palatability of Central Australian Plant Species to Camels*; Technote No. 116. Department of Primary Industry, Fisheries and Resources, Northern Territory Government: Alice Springs, Australia, 2003. Available online: www.nt.gov.au/ d/Content/File/p/Technote/TN116.pdf.

68. Tan, E.T.T.; Al Jassim, R.A.M.; Cawdell-Smith, J.A.; Ossedryver, S.M.; Fletcher, M.T.; The University of Queensland, Brisbane, Queensland, Australia. Unpublished work. 2015.

69. Suliman, H.B.; Wasfi, I.A.; Tartour, G.; Adam, S.E.I. The effects of *Indigofera hochstetteri* on goats. *Rev. Elev. Med. Vet. Pays Trop.* **1983**, *36*, 393–402.

70. Rosenberg, M.M.; Zoebisch, O.C. A chick test for toxicity in forage legumes. *Agron. J.* **1952**,*44*, 315–318.

71. Morris, M.P.; Pagan, C.; Warmke, H.E. Hiptagenic acid, a toxic component of *Indigofera endecaphylla*. *Science* **1954**, *119*, 322–323.

72. Britten, E.J.; Matsumoto, H.; Palafox, A.L. Comparative toxic effects of 3-nitropropionic acid, sodium nitrite and *Indigofera endecaphylla* on chicks. *Agron. J.* **1959**, *51*, 462–464.

73. Britten, E.J.; Palafox, A.L.; Matsumoto, H. Selection for toxicity in single plants of *Indigofera endecaphylla* by biological assay. *Agron. J.* **1959**, *51*, 651–654.

74. Britten, E.J.; Palafox, A.L.; Frodyma, M.M.; Lynd, F.T. Level of 3-nitropropanoic acid in relation to toxicity of *Indigofera spicata* in chicks. *Crop Sci.* **1963**, *3*, 415–416.

# CHAPTER 9

# Effects of spray-dried animal plasma on food intake and apparent nutrient digestibility by cats when added to a wet pet food recipe

**Carmen Rodríguez, Neus Saborido, Jesus Ródenas, Javier Polo***

APC Europe, S.A. Avda. Sant Julià 246-258, Pol. Ind. El Congost, 08403 Granollers, Spaina

## HIGHLIGHTS

- Under these study conditions, no differences in body weight, feces production and fecal score were found between treatments.
- The addition of SDAP to the recipe improved apparent dry matter, crude fiber, ash, calcium and phosphorus digestibility.
- Results suggest that some biologically active components in SDAP are able to survive the sterilization and canning process.
- Results suggest that SDAP may retain biological functions with beneficial effects on the digestive system of adult cats.

## ABSTRACT

Spray dried animal plasma (SDAP) is a preferred binder in canned pet food products due to its high protein content and excellent physicochemical properties and because it is highly preferred by cats. Eleven mixed breed adult cats were used in a crossover digestibility design study to determine apparent nutrient digestibility of canned chunks and gravy food containing either 30 g/kg SDAP or 30 g/kg wheat gluten (WG) added to the emulsion as binders for producing chunks prior to addition of gravy and subsequent sterilization and canning. Cats were fed 400 g/d of each diet for a 7 d acclimation period followed by a 5 d fecal collection period. Cats fed the SDAP diet had improved ($P < 0.05$) apparent dry matter, crude fiber, ash, phosphorus, and calcium digestibility compared to the WG diet. Results indicate that components in SDAP may retain biological functions that exert beneficial effects on the digestive system of healthy adult cats even after sterilization and canning.


## ABBREVIATIONS

SDAP - spray-dried animal plasma

DM - dry matter

CF - crude fiber

CP - crude protein

GE - gross energy

TDF - total dietary fiber

WG - wheat gluten


## KEYWORDS

Apparent digestibility; Cats; Canned pet food; Spray-dried plasma; Wet pet food; Wheat gluten

# 1. INTRODUCTION

Spray-dried animal plasma (SDAP) is a common ingredient that is preferred in canned pet food products due to its high protein concentration and technological properties. Spray-dried animal plasma has excellent water-holding capacity, foaming and emulsifying properties (Tybor et al., 1975, Etheridge et al., 1981, Caldironi and Ockerman, 1982 and Polo et al., 2005) and gel strength when heated above 80 °C (Parés et al., 1998, Polo et al., 2005 and Polo et al., 2007). In addition, because SDAP is an animal protein, strict carnivores such as cats prefer diets formulated with SDAP to those that use wheat gluten (WG) or other vegetable proteins (Polo et al., 2005 and Polo et al., 2007). For the purposes of pet food manufacture, blood is hygienically collected from healthy animals approved for slaughter for human consumption. The liquid plasma is obtained after centrifugation of whole blood (either from bovine or porcine origin) and further concentrated by membranes before being dehydrated using spray-drying technology. The spray-drying technique is a mild dehydration process that maintains the functional physicochemical and biological properties of the product.

Spray-dried animal plasma is commonly used as a functional protein ingredient in diets of farm animals to improve performance and health conditions (Torrallardona, 2010 and Boyer et al., 2015). In addition, SDAP has been demonstrated to improve nutrient digestibility by pigs (Torrallardona, 2010), poultry (Campbell et al., 2003), fish (Gisbert et al., 2015) and dogs (Quigley et al., 2004).

The inclusion of SDAP in dry dog food kibble either before or after extrusion improved dry matter (DM), crude fiber (CF) and total dietary fiber (TDF) digestibility, and decreased fecal DM excretion (Quigley et al., 2004). The extrusion process used in the manufacturing of dry food kibbles and the sterilization process used in the manufacturing of canned food involves high temperature and pressure that may denature major proteins present in SDAP (APC Inc., unpublished data). However, improved digestibility observed in dry kibbles when SDAP was added before extrusion suggested that changes in digestibility were independent of the functional proteins or that certain bioactive components in SDAP survive extrusion.

Therefore, the main purpose of the study was to determine the effect of SDAP when included in a canned pet food recipe on intake and apparent digestibility of major dietary components in adult cats.

# 2. MATERIALS AND METHODS

## 2.1. Ethic statement

The facility of Kennel "De Morgenstond" (Dussen, The Netherlands) was used for the digestibility study. The facility was maintained according to the Dutch regulations (Animals Act, 2011) and the cats were housed in compliance with the Directive 2010/63/EU of the European parliament and of the council of 22 September, 2010 on the protection of animals used for scientific purposes. A consent permission was obtained from the owner of the cats. The animal experimental procedure was in accordance with the Association of American Feed Control Officials guidelines (AAFCO, 2001). The zootechnical procedures used in this experiment (i.e. weighing) were carried out by experienced and authorized personnel and did not cause animal suffering. Technicians at the kennel were unaware of the composition of diets.

## 2.2. Diets

Two experimental recipes of chunks in gravy were produced in the pilot plant of APC-Europe, S.A. (Granollers, Spain) according to the recipes indicated in Table 1. Each diet provided complete and balanced nutrition for the maintenance of adult cats (AAFCO, 2001 and FEDIAF, 2013).

Frozen animal by-products (previously were ground through a 3 mm screen) were mixed with the different additives (salt, sodium polyphosphate, dextrose, mineral premix, vitamin premix) in a pilot plant meat bowl-chopper (Cato S.A., Spain) at maximum speed (2600 rpm). After mixing, poultry meal and then SDAP or WG (used as control binder) was also added to the meat bowl-chopper. The blending process was finished when the raw emulsion reached 11–14 °C. A special designed device was adapted to the piping of a stuffing machine (Cato S.A., Spain) to produce 0.5 cm thick and 1.5 cm wide ropes on stainless steel trays. The ropes were cooked in a wet steam oven (Doleschal, Austria) at 90 °C for 10 min. After cooking, the ropes were left cooling in the refrigerator (4 °C) for 20 min before manual cutting with a knife into small 0.5–1.0 $cm^2$ chunks. Cans of 400 g capacity were filled with 224.6 g of chunks and 175.4 g of gravy. The ratio of chunks:gravy was exactly 56.15:43.85 respectively (Table 1). All cans were sealed and sterilized at 121 °C for 1 h in a laboratory autoclave (Matachana, Spain) and were left to cool at room temperature for at least two weeks before analysis. One can per production batch of each SDAP or WG recipe (4 cans per recipe) was opened and its contents (chunks and gravy) was homogenized with a domestic electric beater (Groupe SEB Ibérica, Barcelona, Spain) to perform analysis for DM, protein, ash, crude fat, CF, calcium and phosphorus.

For the digestibility trial, a total of 306 cans were produced in different daily production batches according to the capacity of production of the pilot plant used; therefore, 153 cans of each SDAP or WG recipe were produced in four different batches.

Cans were labelled as A (SDAP) or B (WG) without any more identification and sent to the animal facility where the trial was done (Kennel "De Morgenstond", The Netherlands). The digestibility trial started on March 9th and finished on April 1st, 2015.

## 2.3. Animals and feeding protocol

Eleven healthy adult mixed breed cats (six females and five males) with an average age of 143.9 ± 10.5 months and 4.08 ± 0.20 kg body weight were used in a crossover digestibility study design. Each period of the experiment included a 7-day adaptation phase and a 5-day fecal collection phase. Cats were fed 400 g/d of diet always at the same time and the food remaining in the bowl was removed and weighed to calculate food intake. Cats were subjected to a natural light environment during the study and the temperature was maintained at 18 ± 2 °C. Cats had continuous access to fresh and clean water.

During the first 7 days of acclimation, all animals receiving the same diet were housed together as a group in playgrounds except during the daily feeding time at 0800 h and 1600 h. From days 8–12, cats were placed in individual cages to collect and to evaluate the production of feces.

All cats were weighed on days 0, 7 and 12 of each experimental period. Daily food consumption was noted per animal during the complete trial.

**Table 1.** Chunk recipe of the two wet pet food products used in the study.

| Ingredients (g/kg)[a] | Recipe A SDAP | Recipe B WG |
|---|---|---|
| Chicken carcass (50:50 fresh:frozen) | 474 | 474 |
| Pig liver | 30.0 | 30.0 |
| Poultry meal | 20.2 | 20.2 |
| Spray-dried plasma | 30.0 | 0.000 |
| Wheat gluten | 0.000 | 30.0 |
| Sodium chloride (NaCl) | 2.40 | 2.40 |
| Sodium tripolyphosphate | 2.00 | 2.00 |
| Dextrose | 1.40 | 1.40 |
| Cat vitamin premix | 0.700 | 0.700 |
| Cat mineral premix | 0.300 | 0.300 |
| Total chunk | 561 | 561 |
| Potassium chloride (KCl) | 1.20 | 1.20 |
| Sodium carbonate ($Na_2CO_3$) | 0.700 | 0.700 |
| Chicken broth (C1301) | 8.00 | 8.00 |
| Water | 429 | 429 |
| Total gravy | 439 | 439 |
| TOTAL | 1000 | 1000 |

[a] Frozen animal by-products (chicken carcass and pork liver) from CORSA Petfood, S.L. (Barcelona, Spain). Fresh poultry carcass from Costa Angelet (Vilanova del Vallès, Barcelona, Spain). Sodium chloride

(NaCl) food grade from Unión Salinera de España (Barcelona, Spain). Sodium tripolyphosphate food grade from Payon (Engis, Belgium). Dextrose from Cargill (Saint Germain en Laye, France). Poultry meal from Avifood (Tarragona, Sain). Potassium chloride from k+s Kali (Kassel, Germany). Sodium carbonate from Quimivita (Barcelona, Spain). Chicken broth from Essentia (Ankeny, US). Spray-dried plasma (AP820P) from APC-Europe (Granollers, Spain). Wheat Gluten Vital from Roquette (Lestrem, France). Chicken broth (C1301) from Essentia Protein Solutions (Ankeny, IA, USA). Vitamin and mineral premixes supplied by Affinity Petcare in Sant Cugat del Vallès (Barcelona, Spain). Vitamin premix contains: 38,400 MIU/kg vitamin A as retinyl acetate; 2560 MIU/kg vitamin D3 as cholecalciferol; 136 g/kg vitamin E as di-alpha-tocopheryl acetate; 24 g/kg vitamin B1 as thiamine mononitrate; 12.8 g/kg vitamin B2 as riboflavin 80% spray-dried; 32 MIU/kg vitamin B5 as calcium pantotenate 98%; 12.8 g/kg vitamin B6 as pyridoxine hydrochloride; 4.8 g/kg vitamin B9 as folic acid 96%; 64 mg/kg vitamin B12 as cianocovalamin on CaCO3 flour; 128 g/kg niacin as nicotinic acid; 112 mg/kg biotin as biotin 2% on CaCO3 flour; 320 mg/kg vitamin K as k3 menadione nicotinamide bisulfite and 45% calcium carbonate (carrier). Mineral premix contains: 8000 mg/kg cooper as cupric sulphate; 1310 mg/kg iodine as potassium iodine; 131 g/kg zinc as zinc sulphate; 100 mg/kg selenium as sodium selenite; 78 g/kg iron as ferrous sulphate; 36.5 g/kg manganese as manganous sulphate and sepiolite or calcium carbonate as carriers.

## 2.4. Samples

Total feces were collected daily during the 5 day collection period, weighed and frozen at −20 °C. Fecal score was evaluated by trained people using the following scale: 0 = no feces visible in the cage at observation time; 1 = liquid diarrhea; 1.5 = diarrhea; 2 = wet, shapeless feces; 2.5 = wet, slightly shaped feces; 3 = wet, distinctly shaped feces; 3.5 = sticky, well-shaped feces; 4 = well shaped feces; 4.5 = hard feces; 5 = hard, very dry feces likely to crumble easily. At the end of each period, all the feces produced by each animal were mixed and freeze-dried (Liolabor 30, Coolvacuum Technologies SL, Pallejà, Spain). After freeze-drying, feces were ground, sieved (discarding hair, straw and foreign materials) and stored at 4 °C until analysis.

## 2.5. Analysis

Fecal samples and experimental cans (chunks mixed homogeneously with gravy) were analyzed externally for dry matter (AOAC method 925.45), crude protein (AOAC method 990.03), ash (AOAC method 942.05), crude fat (AOAC method 954.02), crude fiber (AOAC method 962.09), calcium (AOAC method 968.08) and phosphorus (AOAC method 965.17) (AOAC, 2001). Total tract apparent digestibility of dietary components was determined from these data. Crude fiber (CF) was used as an indicator of fiber digestion; however, use of CF as an index of fiber digestion ignores the loss of soluble polysaccharides, of some insoluble polysaccharides and of lignin (Fahey et al., 1990a and Fahey et al., 1990b). This makes other indices of dietary fiber, such as total dietary fiber (TDF), preferable to CF but unfortunately, TDF was not measured in this study. Gross energy (GE) was estimated by applying the NCR equation (NRC, 2006).

## 2.6. Statistical analysis

Results are expressed as mean of eleven cats. Statistical analysis of the data were performed with the aid of Statgraphics Centurion WX, version 15.2.14 (Manugistics, Inc., USA). The effect of diet and day of study and their interactions, on food consumption and on the production and consistency of feces was done by One-Way ANOVA using LSD as an all pairwise multiple comparison procedure. Differences were considered significant when $P < 0.05$. When comparisons were made between two groups Student *t*-test was used and differences were considered significant when $P < 0.05$.

# 3. RESULTS

## 3.1. Recipes

There were no significant differences in dry matter, calcium, phosphorus and fiber content between diets (Table 2). However, the diet containing SDAP had lower protein, higher fat and higher ash content compared to the diet containing WG.

**Table 2.** Nutrient composition of chunk and gravy recipes[a].

| | Diet A SDAP | Diet B WG | SEM | P-value[b] |
|---|---|---|---|---|
| Dry matter (g/kg) | 268 | 274 | 2.22 | 0.176 |
| Protein (g/kg) | 123 | 135 | 3.23 | 0.012 |
| Ash (g/kg) | 40.3 | 35.9 | 1.27 | 0.054 |
| Calcium (g/kg) | 9.40 | 9.93 | 1.89 | 0.148 |
| Phosphorus (g/kg) | 6.00 | 5.90 | 1.13 | 0.677 |
| Crude fat (g/kg) | 99.3 | 80.5 | 4.47 | 0.003 |
| Crude fiber (g/kg) | 1.90 | 1.71 | 0.285 | 0.749 |

[a] Values are expressed as the average of four different samples in SDAP diet and four different samples in WG diet. Samples were obtained by homogenization of chunks with gravy contained in one can.

[b] Probability of a significant treatment effect by Student *t*-test.

## 3.2. Food intake

The diet had no significant effect on daily food consumption during the whole study. By multi-factorial ANOVA (diet and day) no significant effect was found for diet ($P = 0.529$) or day of trial ($P = 0.998$), and there was no effect related to the interaction of diet and day of trial ($P = 0.958$) (data not shown).

No differences in dry matter, crude protein (CP), CF, ash, calcium, phosphorus or GE were observed between diets; however, cats fed the control diet containing WG had lower ($P = 0.005$) fat ingestion compared with the cats fed the SDAP diet (Table 3).

## 3.3. Feces production and consistency

The daily production of feces during the collection period was not significantly affected by diet ($P = 0.572$) or day of collection ($P = 0.240$), and there was no effect related to the interaction of diet by day ($P = 0.982$) (data not shown). No differences of dry fecal excretion were observed between diets (Table 3).

The fecal consistency was not different during the collection period. No significant effect was found for either outcome considered, diet ($P = 0.515$) and day of trial ($P = 0.823$), and there was no effect related to the interaction of both ($P = 0.798$) (data not shown).

## 3.4. Body weight

There were no significant differences in body weight measured on days 1, 7 and 12 of the assay between diets (Table 3).

## 3.5. Apparent digestibility

Apparent digestibility of CF ($P < 0.001$), ash ($P = 0.001$), calcium ($P = 0.053$), phosphorus ($P = 0.022$) and DM ($P = 0.019$) was improved in the SDAP diet compared to the WG diet (Table 3). Furthermore, the diet containing SDAP had numerically higher digestibility of CP, fat, and GE.

**Table 3.** Mean body weight, fecal excretion, fecal scores, intake and apparent digestibility of cat food diets containing 30 g/kg spray dried animal plasma (SDAP) or wheat gluten (WG)[a].

| Item | SDAP[b] | WG[b] | SEM | *P*-value[c] |
|---|---|---|---|---|
| Body weight, kg | | | | |
| d 1 | 3.87 | 4.13 | 0.141 | 0.368 |
| d7 | 4.00 | 4.02 | 0.157 | 0.954 |
| d 12 | 3.92 | 3.97 | 0.159 | 0.866 |
| Feces, g/d (wet) | 30.0 | 33.6 | 4.50 | 0.569 |
| Feces, g/d (dry) | 5.90 | 7.56 | 0.881 | 0.347 |
| Fecal score[d] | 3.20 | 3.40 | 0.113 | 0.358 |
| Nutrient intake | | | | |
| Intake (g/d) | 240 | 235 | 8.78 | 0.794 |
| Dry matter, g/d | 63.5 | 64.4 | 2.38 | 0.843 |
| CP, g/d | 29.2 | 32.0 | 1.14 | 0.217 |
| CF, g/d | 0.455 | 0.400 | 0.016 | 0.119 |
| Fat, g/d | 23.5 | 18.9 | 0.835 | 0.005 |
| Ash, g/d | 9.55 | 8.45 | 0.342 | 0.106 |
| Calcium, g/d | 2.23 | 2.33 | 0.085 | 0.550 |
| Phosphorus, g/d | 1.42 | 1.39 | 0.052 | 0.752 |
| GE, kcal/d | 391 | 380 | 14.4 | 0.693 |
| Apparent Digestibility, coefficient | | | | |
| DM | 0.926 | 0.890 | 0.008 | 0.019 |
| CP | 0.938 | 0.926 | 0.006 | 0.314 |
| CF | 0.957 | 0.857 | 0.013 | <0.001 |
| Fat | 0.969 | 0.949 | 0.007 | 0.122 |
| Ash | 0.742 | 0.595 | 0.030 | 0.001 |
| Calcium | 0.621 | 0.510 | 0.035 | 0.053 |
| Phosphorus | 0.690 | 0.576 | 0.031 | 0.022 |
| GE | 0.950 | 0.937 | 0.005 | 0.192 |

[a] n = 11 adult cats per treatment.

[b] Values are expressed as the average of 11 values in each group.

[c] Probability of a significant treatment effect by Student *t*-test.

[d] Feces were scored daily on a scale of the following: 0 = no feces visible in the

## 4. DISCUSSION

Spray-dried animal plasma and wheat gluten are protein sources that are widely used in canned pet foods as binder ingredients because of their high capacity to produce a mild gel after heat treatment, and to retain water under hot conditions (Polo et al., 2005 and Polo et al., 2007). Therefore, use of wheat gluten in the control diet was appropriate comparing the digestibility of SDAP against a common binder used in this application.

The diets used in this study contained only three main animal by-products (chicken carcass, pig liver and poultry meal) that accounted for 93% of the chunk recipe (without the gravy) due to the fact that the intention of the study was to test the digestibility of a 3% inclusion level of SDAP versus 3% WG. In addition, because of the small scale chunk production in our facility, 4 days of production for each diet was needed, and although care was taken to homogenize the raw material for each production lot, we could not prevent recipe variation in protein, ash, and fat concentrations. The differences in protein and ash composition of the recipes may be explained by the lower protein and high ash content of SDAP (696 and 147 g/kg, respectively) versus that of WG (767 and 8 g/kg, respectively). The differences in fat content between recipes are related to small differences in fat in the poultry carcass used in the different production batches. This is a common situation in wet pet food industry due to the lack of homogeneity of the raw materials of animal origin (Polo et al., 2007). Nevertheless, the amount of nutrients ingested by cats between the two treatments was similar and only fat ingestion was different. Cats fed WG diet had lower fat ingestion compared to cats fed SDAP diet due to the lower fat content of the WG diet.

Results of this study indicated no differences in body weight, as this was expected due to the ingestion of similar energy intake between treatments. No differences was also observed for wet or dry feces production. Dogs fed dry kibbles containing SDAP included at 1–3% of the diet had no differences in body weight after 15 days; however, there was a consistent reduction in fecal DM output by an average of 15%

across 3 trials (Quigley et al., 2004). We did not observe reduced fecal DM output reported for dogs (Quigley et al., 2004) in our study with cats, although dry feces output was numerically lower for cats fed the SDAP diet.

Fecal score was not different between treatments. This is in agreement with results from the study conducted with dogs (Quigley et al., 2004). However, for other animal species, inclusion of SDAP is well correlated with improved fecal score and reduced diarrhea (Van Dijk et al., 2001, Quigley et al., 2002 and Torrallardona, 2010), probably because animals included in these studies were subjected to a more challenging pathogen environment and the use of dietary SDAP had a greater beneficial effect under these conditions (Coffey and Cromwell, 1995 and Bregendahl et al., 2005). Dietary SDAP may improve the health of challenged animals by different mechanisms, including the antimicrobial protection of immunoglobulins and glycoproteins present in SDAP that may prevent growth and bacterial colonization in the small intestine, support the intestinal immune system, and prevent mucosal damage by pathogenic bacteria (Moretó and Pérez-Bosque, 2009). In addition, SDAP has been reported to reduce the overall activation of the immune system and have a systemic anti-inflammatory effect (Moretó and Pérez-Bosque, 2009 and Maijó et al., 2012).

Digestibility results were in the upper range of values reported in the literature for cats fed canned pet food (Nott et al., 1994, Kerr et al., 2013 and Kerr et al., 2014). This may be due to the relatively simple diet composition and the high quality animal by-products used in our study. In fact, our values are similar to values obtained when cats were fed raw meat products (Kerr et al., 2012). We also observed higher digestibility for all nutrients measured compared with dogs fed a dry kibble containing SDAP from 1 to 3% (Quigley et al., 2004), probably because we are comparing two different species and two different processing methods (extrusion vs retorting).

In our study, we observed higher digestibility of the SDAP diet for DM, CF, ash, calcium, and phosphorus. Higher numerical digestibility values were observed for CP, crude fat, and GE for the SDAP diet. Due to the presence of IgG in SDAP, which is partially resistant to digestion, the digestibility of crude protein in diet containing SDAP was expected to be slightly lower (Roos et al., 1995 and Quigley et al., 2004). Bosi et al. (2001) reported lower ileal CP digestibility in early-weaned pigs fed SDAP compared with hydrolyzed casein. Conversely, Pendergraft et al. (1993) reported that SDAP was more digestible than WG in diets fed to young pigs. Similarly, Rojas and Stein (2012) demonstrated that SDAP had higher CP and amino acid digestibility values than common animal proteins used in pig diets (fishmeal, casein, chicken meal, etc.), however, as indicated above, in our study no difference on CP was observed between treatments.

Our results indicating higher digestibility of CF in the diet containing SDAP is in agreement with the results obtained for dogs (Quigley et al., 2004). Because SDAP is not a significant source of fiber, we were not expecting differences in fiber digestion, however, the higher digestibility of crude fiber in SDAP diets may be related with improved intestinal microbiota that efficiently use the fiber provided in the diet. Torrallardona et al. (2003)indicated higher counts of lactobacillus in the ileum and cecum of piglets fed a diet containing 7% SDAP.

The higher digestibility of phosphorus as a result of the inclusion of SDAP is in agreement with the high digestibility of phosphorus observed in pigs fed diets containing SDAP (Almeida and Stein, 2011). Similarly, we observed a higher digestibility of calcium and total ash. These results indicated that minerals present in SDAP are very digestible and soluble and, consequently, highly available for the animals. Nevertheless, the contribution of phosphorus and calcium from the inclusion of SDAP at 30 g/kg is minimal and therefore the better digestibility observed by the addition of SDAP in the wet pet food recipes can be explained by the improved gastrointestinal health associated with the supplementation with this ingredient (Quigley et al., 2004, Pérez-Bosque et al., 2006 and Peace et al., 2011). This can be an important observation for diets designed for all ages, but especially for adult pets.

Our results indicated that functional components present in SDAP survived canning conditions and improved digestibility of the diet. This observation is in agreement with the data published by Quigley et al. (2004) who observed that SDAP added in dry kibbles before extrusion improved digestibility of select nutrients. In addition, Campbell et al. (2006) demonstrated that poultry diets containing 1% SDAP and subjected to pellet conditioning temperatures ranging from 85 to 95 °C and expander temperatures of 149 °C did not impair the positive growth effects of SDP in pelleted or expanded broiler feed. Similarly, Gao et al. (2011) fed 10% SDAP or 10% autoclaved SDAP (auSDAP) in diets of neonatal pigs and observed that both SDAP and auSDAP improved the growth performance of neonatal pigs, although the SDAP diet had a greater effect on growth rate. According to the authors, the benefits of SDAP probably resulted from the promotion of intestinal development, accompanied by increased antioxidant capacity and decreased production of inflammatory factors in the intestinal mucosa. Conversely,Steidinger et al. (2000) demonstrated that feeding diets containing 5% SDAP and pelleted at conditioning temperatures above 77 °C decreased weanling pig growth performance. However, due to the presence of reducing sugars from lactose and starch commonly using in weaned pig diets, the reduced growth performance results observed for pelleted pig diets conditioned at greater than 77 °C was likely the result of Maillard reaction product formation that reduced the amino acid digestibility of the diet. Our results with SDAP in recipe emulsions that were subsequently cooked at 90 °C for 10 min and sterilized at 121 °C for 1 h, suggest that the improved digestibility of select nutrients was independent of the functionality of proteins, or that functional components in SDAP survive sterilization, or that some bioactive components are produced after digestion of the plasma proteins.

Plasma is a complex mixture of many physiological components including immunoglobulins, glycoproteins, growth factors, bioactive peptides, cytokines and others (Coffey and Cromwell, 2001 and Moretó and Pérez-Bosque, 2009). All these components provide anti-bacterial (Nollet et al., 1999 and Niewold, 2007) and anti-inflammatory effects (Pérez-Bosque et al., 2004, Pettigrew et al., 2006 and Peace et al., 2011), improve intestinal barrier function (Pérez-Bosque et al., 2006 and Peace et al., 2011;), and have other beneficial effects such as increase disease resistance, modulation of the immune system. It appears to dampen inflammation in healthy animals, and that effect may be important in increasing both feed intake and growth rate. In contrast, plasma appears to enhance the immune response to a challenge. (Pettigrew et al., 2006).

Overall, these data suggest that SDAP may influence gastrointestinal system function, and thereby enhance digestion. Addition of SDAP to canned cat food improved digestibility of nutrients. Improved digestibility is consistent with better digestive function observed in other animals when SDAP was included in the diet and may indicate that other components present in SDAP are resistant to high pressure and thermal processing conditions and are responsible for some of the effects observed with the supplementation of SDAP in animal diets.

# 5. CONCLUSIONS

SDAP in canned chunk and gravy food for cats improved apparent digestibility of dry matter, crude fiber, ash, calcium, and phosphorus. These results are similar to those reported for dry kibbles containing

SDAP in digestibility studies with dogs (Quigley et al., 2004). In addition to the well-known superior physicochemical properties of SDAP as a binder, functional components in SDAP apparently retain at least some of its biological activity through high pressure and thermal processing conditions to enhance digestibility in both canned and dry extruded recipes.

## CONFLICT OF INTEREST

The authors have read the journal's policy and the authors of this manuscript have the following competing interests: CR, NS, JR and JP are employed by APC Europe, S.A., Granollers, Spain. This does not alter our adherence to all the Animal Feed Science and Technology policies on sharing data and materials. APC Europe, S.A. manufacture and sells spray dried plasma.

## ACKNOWLEDGEMENTS

We want to thank all staff of Kennel "De Morgenstond", especially to all directly involved in taking care of the cats. In particular, we are deeply indebted to Martin Hoogerwaard, the General Manager of Kennel "De Morgenstond", for all his help and advice provided during this study.

## REFERENCES

1. AAFCO, 2001. Official Publication. Association of American Feed Control Officials Inc., Atlanta, GA, USA.

2. AOAC, 2001. Official Methods of Analysis, 16th ed. Association of Official Analytical Chemists, Washington, DC, USA.

3. Almeida, F.N., Stein, H.H., 2011. Standardized total tract digestibility of phosphorus in blood products fed to weanling pigs. Rev. Colomb. Cienc. Pecu. 24, 617–622.

4. Anonymous. Act of May 19, 2011, constituting an integral framework for regulations on captive animals and related topics (Animals Act). The Minister of Security and Justice. Official Diary of the Kingdom of the Netherlands. Gazette 345. 1–63.

5. Bosi, P., Han, I.K., Jung, H.J., Heo, J.N., Perini, S., Castellazzi, A.M., Casini, L., Creston, D., Gremokolini, C., 2001. Effect of different spray dried plasmas on growth ileal digestibility, nutrient deposition, immunity and health of early-weaned pigs challenged with E. coli K88. Asian Aust. J. Anim. Sci. 14, 1138–1143.

6. Boyer, P.E., D'Costa, S., Edwards, L.L., Milloway, M., Susick, E., Borst, L.B., Thakur, S., Campbell, J.M., Crenshaw, J.D., Polo, J., Moeser, A.J., 2015. Early-life dietary spray-dried plasma influences immunological and intestinal injury responses to later-life Salmonella typhimurium challenge. Br. J. Nutr. 113 (5), 783–793.

7. Bregendahl, K., Ahn, D.U., Trampel, D.W., Campbell, J.M., 2005. Effects of dietary spray-dried plasma protein on broiler growth performance and breast meat yield. J. Appl. Poult. Res. 14, 560–568.

8. Caldironi, H., Ockerman, H., 1982. Incorporation of blood proteins into sausage. J. Food Sci. 47, 405–408.

9. Campbell, J.M., Quigley III, J.D., Russell, L.E., Kidd, M.T., 2003. Effect of spray-dried bovine serum on intake, health, and growth of broilers housed in different environments. J. Anim. Sci. 81, 2776–2782.

10. Campbell, J.M., Russell, L.E., Crenshaw, J.D., Behnke, K.C., Clark, P.M., 2006. Growth response of broilers to spray-dried plasma in pelleted or expanded feed processed at high temperature. J. Anim. Sci. 84, 2501–2508.

11. Coffey, R.D., Cromwell, G.L., 1995. The impact of the environment and antimicrobial agents on the growth response of early-weaned pigs to spray-dried porcine plasma. J. Anim. Sci. 75, 2532–2539.

12. Coffey, R.D., Cromwell, G.L., 2001. Use of spray-dried animal plasma in diets for weanling pigs. Pig News Inf. 22, 39N–48N.

13. Etheridge, P., Hickson, D., Young, C., Landmann, W., Dill, C., 1981. Functional and chemical characteristics of bovine plasma

proteins isolated as a metaphosphate complex. J. Food Sci. 46, 1782–1784.

14. European Pet Food Industry (FEDIAF), 2013. Nutritional Guidelines for Complete and Complementary Pet Food for Cats and Dogs, http://www.fediaf.org/self-regulation/nutrition/ (accessed 14.10.15).

15. Fahey Jr., G.C., Merchen, N.R., Corbin, J.E., Hamilton, A.K., Serbe, K.A., Lewis, S.M., Hirakawa, D.A., 1990a. Dietary fiber for dogs: I. Effects of graded levels of dietary beet pulp on nutrient intake digestibility, metabolizable energy and digesta mean retention time. J. Anim. Sci. 68, 4221–4228.

16. Fahey Jr., G.C., Merchen, N.R., Corbin, J.E., Hamilton, A.K., Serbe, K.A., Lewis, S.M., Hirakawa, D.A., 1990b. Dietary fiber for dogs: II. Isototal dietary fiber (TDF) additions of divergent fiber sources to dog diets and their effects on nutrient intake digestibility, metabolizable energy and digesta mean retention time. J. Anim. Sci. 68, 4229–4235.

17. Gao, Y.Y., Jiang, Z.Y., Lin, Y.C., Zheng, C.T., Zhou, G.L., Chen, F., 2011. Effects of spray-dried animal plasma on serous and intestinal redox status and cytokines of neonatal piglets. J. Anim. Sci. 89 (1), 150–157.

18. Gisbert, E., Skalli, A., Campbell, J., Solovyev, M.M., Rodríguez, C., Dias, J., Polo, J., 2015. Spray-dried plasma promotes growth, modulates the activity of antioxidant defenses, and enhances the immune status of gilthead sea bream (Sparus aurata) fingerlings. J. Anim. Sci. 93, 278–286.

19. Kerr, K.R., Vester Boler, B.M., Morris, C.L., Liu, K.J., Swanson, K.S., 2012. Apparent total tract energy and macronutrient digestibility and fecal fermentative end-product concentrations of domestic cats fed extruded, raw beef-based, and cooked beef-based diets. J. Anim. Sci. 90, 515–522.

20. Kerr, K.R., Morris, C.L., Burke, S.L., Swanson, K.S., 2013. Apparent total tract macronutrient and energy digestibility of 1- to-3-day-old whole chicks, adult ground chicken, and extruded and canned chicken-based diets in African wildcats (Felis silvestris lybica). Zoo Biol. 32 (5), 510–517.

21. Kerr, K.R., Morris, C.L., Burke, S.L., Swanson, K.S., 2014. Apparent total tract energy and macronutrient digestibility of one- to three-day-old, adult ground, extruded, and canned chicken-based diets in domestic cats (Felis silvestris catus). J. Anim. Sci. 92, 3441–3448.

22. Maijó, M., Miró, L., Polo, J., Campbell, J., Russell, L., Crenshaw, J., Weaver, E., Moretó, M., Pérez-Bosque, A., 2012. Dietary plasma proteins attenuate the innate immunity response in a mouse model of acute lung injury. Br. J. Nutr. 107 (6), 867–875.

23. Moretó, M., Pérez-Bosque, A., 2009. Dietary plasma proteins, the intestinal immune system, and the barrier function of the intestinal mucosa. J. Anim. Sci. 87, E92–E100.

24. NRC, 2006. Nutrient Requirements of Dogs and Cats. Ad Hoc Committee on Dog and Cat Nutrition. Natl. Acad. Sci. Press, Washington, DC. USA.

25. Niewold, T.A., 2007. The nonantibiotic anti-inflammatory effect of antimicrobial growth promoters, the real mode of action? A hypothesis. Poult. Sci. 86, 605–609.

26. Nollet, H., Deprez, P., van Driessche, E., Muylle, E., 1999. Protection of just weaned pigs against infection with F18+ Escherichia coli by non-immune plasma powder. Vet. Microbiol. 65, 37–45.

27. Nott, H.M.R., Rigby, S.I., Johnson, J.V., Bailey, S.J., Burger, I.H., 1994. Design of digestibility trials for dogs and cats. J. Nutr. 124, 2582S–2583S.

28. Pérez-Bosque, A., Pelegrí, C., Vicario, M., Castell, M., Russell, L., Campbell, J.M., Quigely III, J.D., Polo, J., Amat, C., Moretó, M., 2004. Dietary plasma protein affects the immune response of weaned rats challenged with S. aureus superantigen B. J. Nutr. 134, 2667–2672.

29. Pérez-Bosque, A., Amat, C., Polo, J., Campbell, J.M., Crenshaw, J., Russell, L., Moretó, M., 2006. Spray-dried animal plasma prevents the effects of Staphylococcus aureus enterotoxin B on intestinal barrier function in weaned rats. J. Nutr. 136, 2838–2843.

30. Parés, D., Saguer, E., Saurina, J., Sunol, ˜ J.J., Carretero, C., 1998. Functional properties of heat induced gels from liquid and spray-dried porcine blood plasma as influenced by pH. J. Food Sci. 63, 958–961.

31. Peace, R.M., Campbell, J., Polo, J., Crenshaw, J., Russell, L., Moeser, A., 2011. Spray-dried porcine plasma influences intestinal barrier function, inflammation, and diarrhea in weaned pigs. J. Nutr. 141, 1312–1317.

32. Pendergraft, J.S., Hancock, J.D., Hines, R.H., Mills, C.G., Burnham, L.L., 1993. Effects of wheat gluten and plasma protein on growth performance and digestibility of nutrients in nursery pigs. In: Swine Day Report. Kansas State Univ., Manhattan, pp. 58–62.

33. Pettigrew, J.E., Gaskins, H.R., Nava, G., 2006. A critical review of functional animal proteins. In: Research Report No: NPB #04-142 in Animal Science. National Pork Board, Available: http://www.pork.org/ FileLibrary/ ResearchDocuments/04-142-PETTIGREW-UofILL.pdf (accessed 10.11.15).

34. Polo, J., Rodríguez, C., Saborido, N., Ródenas, J., 2005. Functional properties of spray-dried animal plasma in canned petfood. Anim. Feed Sci. Technol. 122, 331–343.

35. Polo, J., Rodríguez, C., Ródenas, J., Morera, S., Saborido, N., 2007. Use of spray-dried animal plasma in canned chunk recipes containing excess of added water or poultry fat. Anim. Feed Sci. Technol. 133, 309–319.

36. Quigley, J.D., Kost, C.J., Anspach, T.M., 2002. Effects of oral immunoglobulins and oligosachharides in calf milk replacer formulations. J. Dairy Sci. 85, 413–421.

37. Quigley, J.D., Campbell, J.M., Polo, J., Russell, L.E., 2004. Effects of spray-dried animal plasma on intake and apparent digestibility in dogs. J. Anim. Sci. 82, 1685–1692.

38. Rojas, O.J., Stein, H.H., 2012. Nutritional value of animal proteins fed to pigs nutritional value of animal proteins fed to pigs. In: Proc. Midwest Swine Nutr. Conf., Indianapolis, IN, pp. 9–24.

39. Roos, N., Mahe, S., Benamouzig, R., Sick, H., Rautureau, J., Tome, D., 1995. 15N-labelled immunoglobulins from bovine colostrum are

partially resistant to digestion in human intestine. J. Nutr. 125, 1238–1244.

40. Steidinger, M.U., Goodband, R.D., Tokach, M.D., Dritz, S.S., Nelssen, J.L., McKinney, L.J., Borg, B.S., Campbell, J.M., 2000. Effects of pelleting and pellet conditioning temperatures on weanling pig performance. J. Anim. Sci. 78, 3014–3018.

41. Torrallardona, D., Conde, M.R., Badiola, I., Polo, J., Brufau, J., 2003. Effect of fishmeal replacement with spray-dried animal plasma and colistin on intestinal structure, intestinal microbiology, and performance of weanling pigs challenged with Escherichia coli K99. J. Anim. Sci. 81, 1220–1226.

42. Torrallardona, D., 2010. Spray dried animal plasma as an alternative to antibiotics in weanling pigs—a review. Asian Aust. J. Anim. Sci. 23, 131–148.

43. Tybor, P., Dill, C., Landmann, W., 1975. Functional properties of proteins isolated from bovine by a continuous pilot process. J. Food Sci. 40, 155–159.

44. Van Dijk, A.J., Everts, H., Nabuuers, M.J.A., Margry, R.J.C.F., Beynen, A.C., 2001. Growth performance of weanling pigs fed spray-dried animal plasma: a review. Livest. Prod. Sci. 68, 263–274.

# CHAPTER 10

# Dietary fiber and protein fermentation in the intestine of swine and their interactive effects on gut health and on the environment: A review

**Rajesh Jha*, Julio F.D. Berrocoso**

Department of Human Nutrition, Food and Animal Sciences, College of Tropical Agriculture and Human Resources, University of Hawaii at Manoa, Honolulu, HI 96822, USA

## HIGHLIGHTS

- We reviewed interactive effects of fiber and protein fermentation in the pig gut.
- Fiber fermentation may improve gut health of pig.
- Protein fermentation negatively affects both gut health and the environment.
- Negative effects of protein fermentation can be minimized by dietary manipulation.
- Inclusion of fiber and reduction in protein level in diet can be a better option.

# ABSTRACT

Fermentation of both dietary fiber (DF) and protein in the pig intestine is a matter of interest because of their potential beneficial or harmful effects on gut health and on the environment. This paper reviews some of the relevant information available on DF and protein fermentation and their interactive effects on the gut environment of pigs and its contribution to emission of nitrogenous gases and odor from pig manure and piggeries. The fermentation of protein and the associated production of metabolic compounds are discussed in relation to DF fermentation, their impact on gut health, bacterial protein synthesis and nitrogen (N) excretion. Some nutritional strategies to reduce protein fermentation in the gut such as the reduction of the amount of crude protein (CP) in the diet and/or the inclusion of fermentable DF are also presented. Also, to cope with the negative impact of intensive pig production on the environment, different nutritional approaches such as reducing N excretion by lowering CP intake, shifting the N excretion pathway from urine to feces and lowering the pH of manure by lowering the pH of urine and feces have been reviewed. Overall, inclusion of DF and reduction of CP in pig diets seems to be an effective nutritional strategy that may counteract the negative effects of protein fermentation in the pig gut by reducing ammonia concentration, shifting N excretion pathways in the gut and minimizing the negative impact of intensive pig production on the environment.



# KEYWORDS

Bacterial protein synthesis; Crude protein; Dietary fiber; Gut health; Nitrogen excretion; Pig


# ABBREVIATIONS

AA - amino acid

BCFA - branched chain fatty acids

CHO - carbohydrates

CP - crude protein

DF - dietary fiber

GIT - gastrointestinal tract

N - nitrogen

NSP - non-starch polysaccharides

PWD - post-weaning diarrhea

SBP - sugar beet pulp

SCFA - short chain fatty acid

RS - resistant starch

WBC - water binding capacity

# 1. INTRODUCTION

The modern pig production systems are criticized for their negative impact on the environment due to nitrogenous gases excretion and odor emission from barns. Nitrogenous gases emitted from piggeries are also found to have a negative impact on the health of barn workers as well as the pigs themselves. Even low levels of ammonia (25 ppm) in pig barns induce nasal irritation and depression of growth in pigs (Urbain et al., 1994), while air quality in the confinement of a pig facility can cause acute respiratory responses in humans (Zhang et al., 1998). Thus, measures to mitigate nitrogenous gases excretion and odor emission from pig facilities need attention. To deal with these challenges, several alternative approaches have been forwarded, including nutritional manipulation in the swine feeding program. The effectiveness of these nutritional strategies in reducing nitrogenous gases and odor emission from piggeries and improving gut health and bacterial protein synthesis are based on the effects of diet composition, especially dietary fiber (DF) and protein, on nutrient digestibility and metabolism, and on the fermentation rates in the large intestine.

Ammonia emission from pig manure originates mainly from urea in the urine. Nitrogen (N) in the feces comprises undigested dietary N and endogenous N, mainly as amino acids (AA), and microbial N, partly present in nucleic acids. One approach to reducing ammonia emission is through dietary manipulation. In this respect, DF plays an important role in reducing ammonia emission (Aarnink and Verstegen, 2007) and improving gut health (Williams et al., 2001 and Montagne et al., 2003). Dietary fiber is fermented by gut bacteria, producing short chain fatty acids (SCFA, Jha et al., 2010 and Jha and Leterme, 2012), which in turn, supports the growth of beneficial bacteria and suppress harmful bacteria in the gut (Pieper et al., 2008). The gut bacteria induce a shift of N excretion from urea in urine to feces (Zervas and Zijlstra, 2002 and Bindelle et al., 2009). On the other hand, the urea concentration in manure is highly dependent on protein nutrition and can be altered by changing dietary protein content. A reduction in dietary crude protein (CP), while balancing for AA, decreases the urinary and total N excretion (Zervas and Zijlstra, 2002) without affecting the performance and carcass quality of growing pigs (Jha et al., 2013 and Morales et al., 2015). Moreover, unpleasant odor from stored pig manure results largely from compounds like sulfurous compounds, indolic and phenolic compounds, volatile fatty acids, and ammonia and volatile amines, which are the products of incomplete anaerobic digestion of protein and carbohydrate (CHO). Odors are mainly produced by the microbial conversion of feed components in the large intestine of pigs and after excretion by microbial conversion of excreta under anaerobic conditions in manure. It is generally accepted that odorous compounds produced by protein fermentation in the gastrointestinal tract (GIT) and in manure have a higher offensive sensory response and a higher magnitude of odor sensation than straight-chain volatile fatty acids produced from fermentable CHO. Therefore, CP is the first dietary component that should be altered to minimize odor.

To address these concerns and give a better understanding of interactive effects of DF and protein fermentation and their effects on gut health and on the environment, this paper has reviewed different aspects of DF and CP in swine nutrition. More specifically, DF and protein fermentation, and their pathways and end metabolic products are first briefly reviewed. Also, the nutritional approaches to reduce protein fermentation, the relation between CHO [both fiber and resistant starch, (RS)] and protein fermentation and its effects on gut health, bacterial protein synthesis as well as odor emission from pig manure have been highlighted. Finally, the impact on nitrogenous gases emission from piggeries and the strategies to reduce N excretion and odor emission are also assessed.

# 2. PHYSIOLOGY AND BIOCHEMISTRY OF CARBOHYDRATES AND PROTEIN IN PIGS

The processes in the GIT of pigs are of great importance in understanding the influence of nutrition on metabolism. Microorganisms come in contact with undigested material or endogenous substrate to a considerable degree mainly in the lower parts of GIT, from the terminal ileum to the rectum. The GIT of the pigs can be divided broadly into three different compartments: stomach, small intestine and large intestine. The digestion start in the stomach, which has a lower pH, and a limited amount of microbial activity, may occur in the small intestine when the pig ingests high amounts of feed. Most of the available nutrients (CHO, protein, fat, minerals and vitamins) are absorbed in the small intestine. In the large intestine, undigested feed components (mainly DF, lipids and insoluble protein) and endogenous secretions are fermented by microorganisms. In this part, only SCFA and some vitamins can be absorbed, which contribute to the nutrient supply of the pigs.

Digestion is the process of chemical breakdown that allows absorption of nutrients by enzymes secreted into the lumen of the GIT (Tso and Crissinger, 2000). The enzymes are secreted by glandular cells in the mouth, chief cells in the stomach, exocrine cells in the pancreas, and brush-border intestinal glands (Johnson, 2001). Mammalian enzymes may hydrolyze a limited number of linkages such as α(1–4) in starch and maltooligosaccharides, α(1–6) in starch and dextrins, β(1–2) in sucrose, and β(1–4) in lactose. Other linkages (e.g., β(1–4) in cellulose) are not hydrolyzed by endogenous mammalian enzymes and need to be hydrolyzed by bacterial enzymes in the process of fermentation (Tso and Crissinger, 2000).

The fermentation is an energy-conservation process, in which electrons from a few reactions such as redox, hydrolysis, phosphorylation or lyses are transferred to part of the substrate. From this process, energy is derived, leading to the production of ATP, which is used for bacterial basal and growth metabolism. During fermentation of DF in the pig intestine, the microbes start breaking down polysaccharides into smaller polysaccharides or the constituent monosaccharaides; but in these processes, the substrate is only partially oxidized and only a small amount of energy is extracted for microbial growth. The monomers are absorbed into the microbial cell and channeled into the pathways of central metabolism. The majority of anaerobic microorganisms of the large intestine, except *bifidobacteria*, use the glycogenesis pathway to degrade glucose to pyruvate via glucose-6-phosphate, which is later oxidized to acetate, propionate or butyrate. Pentose and pectin are first metabolized by the pentose phosphate pathway ( Macfarlane and Macfarlane, 2003) starting from the pentose to fructose-6-phosphate and glyceraldehyde-3-phosphate via xylulose-5-phosphate. As a result, SCFA (acetate, propionate and butyrate) and gases ($CO_2$, $H_2$ and $CH_4$) are the main end products of intestinal DF fermentation. Other metabolites such as lactate, ethanol and succinate are also formed by different types of bacteria (Drochner et al., 2004). With the possible exceptions of ethanol, these products do not accumulate in a healthy gut, because they serve as substrate and electron donors for cross-feeding bacteria and are further converted into SCFA (Macfarlane and Gibson, 1995). Further details on fiber fermentation and its effect on the pig intestinal physiology and health can be found in recent review papers ( Aumiller et al., 2015 and Jha and Berrocoso, 2015).

Proteins available to bacteria for fermentation are dietary proteins that escape digestion in the small intestine, along with some proteins of endogenous origin (Macfarlane et al., 1992). The end products of protein fermentation are also different from those of DF fermentation. The proteolytic fermentation results in the production of SCFA, especially branched chain fatty acids (BCFA, mainly iso-butyrate, valerate and iso-valerate). These BCFAs are formed by the metabolism of branched chain AA such as valine, leucine and isoleucine. Among the fermented proteins, about 30% are converted to SCFA, of which BCFA constitute between 16 and 23%, depending on the substrate type (Macfarlane et al., 1992 and Cone et al., 2005). In addition, some potentially toxic metabolites like ammonia and amines (Cone et al., 2005) as well as malodorous compounds such as skatole and indole (Jensen et al., 1995) are also produced. The indolic, phenolic and sulfurous compounds in manure are produced from protein and come from three main pathways (Le et al., 2008). First, protein is partly fermented in the large intestine and the bacterial proteolysis induces the production of BCFA (mainly iso-butyrate, valerate, and iso-valerate), malodorous compounds, amines and ammonia originating from the deamination of AA and indolic and phenolic and sulfurous compounds. These compounds are then transferred to the liver, where these are detoxified to glucuronides and then excreted via urine. Second, unfermented protein in the feces can be fermented in the manure, and indolic, phenolic and sulfurous compounds are produced. Third, a small proportion of indolic, phenolic and sulfurous compounds are excreted directly via feces. In addition, sulfurous compounds in the manure can be produced from sulfates excreted via urine, which mainly originate from excreta of excess sulfur containing AA absorbed in the small intestine of the pig.

# 3. EFFECTS OF MODULATING THE PROTEIN LEVEL IN DIETS

Lowering the amount of CP in the diet (Htoo et al., 2007) is found to be effective in reducing protein fermentation in the gut. The fermentation of undigested dietary protein and proteins of endogenous origin entering the large intestine yields putatively toxic metabolites that can impair epithelial integrity and promote enteric disorder such as the incidence of post-weaning diarrhea (PWD). Most of the published work on piglets (Nollet et al., 1999, Htoo et al., 2007 and Wellock et al., 2008) shows that feeding a low-protein diet after weaning decreased intestinal flow of N and protein fermentation indices in the GIT and that this is linked to a reduction in the incidence of PWD. In this respect, Heo et al. (2010) compared a high-protein diet that contained 240 g/kg CP with a low protein diet with 190 g/kg of CP and reported an increased N flow at the terminal ileum which caused a higher amount of N to be available for bacterial fermentation in the distal gut of the piglets at 35 d of age. Also, increases in microbial protein in the colon should be expected when a sufficient amount of fermentable CHO are present. Also, Htoo et al. (2007)reported that the protein content of pig starter diets can be reduced safely by about 200 g/kg while balancing with limiting AA according to ideal protein ratios, which will also reduce the production of potentially harmful microbial metabolites in the caecum.

## 3.1. Changes in the gut microbiota and fermentation end products that alter the gut health

Dietary and endogenous proteins are utilized in the GIT through microbial fermentation. Bacterial AA metabolism occurs via oxidative and reductive reactions including deamination and decarboxylation. A large variety of bacterial species such as*Escherichia coli*, *Klebsiella* spp., *Campylobacter* spp., *Streptococcus* spp., *Clostridium perfringens*, *Clostridium difficile* and *Bacteroides fragilis* have been reported as dominant protein fermenters. Bacterial AA utilization also contributes to the formation of SCFA in the GIT. Protein fermentation products such as ammonia can interfere with the oxidative metabolism of SCFA in colonocytes, likely inducing energy deficiency in the cell ( Blachier et al., 2007). Decarboxylation of AA yields several biogenic amines such as histamine that can induce chloride secretion and diarrhea in the colon of pigs (Kröger et al., 2013).

Among the several factors responsible for diarrhea in pigs, enterotoxigenic strains of *E. coli* are major ones ( Williams et al., 2001). The *E. coli* colonize the small intestine under multifactorial conditions and a high level of dietary protein predisposes to the condition.Heo et al. (2009) found that pigs challenged with *E. coli* had lower growth rate and decreased gain-to-feed ratio compared with non-challenged pigs from 28 to 70 d of age. However, there was a marked reduction in the incidence of diarrhea after infection with β-hemolytic *E. coli* and feeding a low-protein diet for 7 or 14 d after weaning. It suggests that protein fermentation selectively affects the microbiota in the gut and can influence the pig's health condition.

Increase in protein fermentation leads to increased ammonia and amine concentration in the colon (Macfarlane et al., 1992), whereas these compounds are found only in small amounts in a healthy colon. The increased ammonia concentration in the gut can negatively affect the development of the intestinal mucosa and villus height (Nousiainen, 1991 and Davila et al., 2013), which in turn adversely affect the digestion and absorption processes in the intestine. Ammonia generated in the colon readily passes across the gut wall; thereby gaining access to other tissues of the body, which can be detrimental for the host's health (Cone et al., 2005 and Bikker et al., 2006).

## 3.2. Changes in the N emission in feces or urine and gases

The main nutrients of environmental concern are N, phosphorus and heavy metals, while the main gaseous losses are ammonia and methane (Aarnink and Verstegen, 2007 and Beccacia et al., 2015). Most of the ammonia in pig manure originates from the breakdown of urea while a small amount comes from the breakdown of protein in feces (Aarnink et al., 1993). The activity of urease determines the rate at which urea is converted into ammonia. Urease is present only in feces, not in urine. Thus, the conversion of urea to ammonia only starts when urine mixes with feces or comes in contact with soiled floors (Aarnink et al., 1997). At the manure storage pit, protein breakdown from manure is a slow process, which might take weeks or even longer. Is it assumed that almost half of the N excreted by urine and feces can emit during storage of the manure inside the pigs' house and during surface application of the manure. On the other hand, the degradation of urea to ammonia and carbon dioxide can occur only in few hours, but the process is affected by several factors like ammonia concentration, pH, temperature, air velocity and emitting surface area (Aarnink and Verstegen, 2007).

To deal with the negative impact of intensive pig production on the environment, nutritional approaches have been suggested, such as reducing N excretion by lowering CP intake, shifting N excretion from urea in urine to protein in feces, and lowering the pH of manure by lowering the pH of urine and feces (Aarnink and Verstegen, 2007). The concentration of urea in urine and the pH of feces and urine are important characteristics of excreta to determine ammonia emission from a pig facility. The urea concentration of urine highly depends on the protein level of the diet (Jongbloed, 2008). Manipulating the protein content can thus change it. In practice, dietary protein levels generally exceed the pig's requirements and are thus not totally utilized by the pigs. Several studies have shown that the protein content of diet can be reduced by 3–4% without any negative effect on growing pig performance (Canh et al., 1998b, Ball and Möhn, 2003 and Jha et al., 2013). Some portions of proteins, for example ~25% of protein in a typical corn-soybean meal based diet, cannot be utilized, due to unbalanced AA. These overabundant or unbalanced proteins/AA are broken down to N and excreted as urea in urine. Thus, supplying balanced protein/AA contents and matching the closest possible amount to the requirement of the pig can contribute to reduce N excretion.

Using a mathematical model, Aarnink et al. (1993) estimated that the reduction of CP by 10 g/kg in a diet can reduce ammonium N in manure by 9%, which is almost double that of the first documented report on the role of reduced CP level on reducing N excretion (Lenis, 1989). Confirming the model estimate, Mohn and Susenbeth (1995) found that a reduction of 200 g/kg in dietary protein can reduce N excretion by up to 35%, if the low-protein diets offer adequate AA concentration. Thereafter, several attempts were made to explore the in-depth relationship of CP content in diet and N excretion. Canh et al. (1998b), in comparing three levels of protein content (165, 145 and 125 g/kg), found that every 10 g/kg reduction in CP in diet reduced ammonia content of the manure by 10% and ammonia emission by 10–12%. The effect of CP and the ammonia reduction ratio was a little higher in this study than in studies reported by other researchers, which might be explained by the fact that the pH of the manure had been lowered. Similarly, Sutton et al. (1999) reported a 28% reduction of ammonium and total N concentration in manure when CP was reduced by 30 g/kg (from 130 to 100 g/kg) in corn–soybean meal-based diets supplemented with lysine, tryptophan, threonine and methionine in growing pigs. Such reduction in N can be expected from sows as well (Ball and Möhn, 2003). Reviewing several studies, Kerr (1995) noted that there is wide variation in the effect of reducing the CP content with AA supplementation on N excretion, ranging from 3.2 to 62% (average around 8.4%), which depends on the size of the pig, level of dietary CP reduction, and initial CP level in the control diet.

### 3.3. Changes in odorous compounds

The reduction of dietary CP in pig's diets and supplemental synthetic AA decreases manure concentrations of odorous compounds and maintains animal performance. In this respect, Hobbs et al. (1996) compared a control diet with 200 g/kg CP with a diet that contained with 140 g/kg CP and reported that reducing dietary CP but providing essential AA in an ideal protein ratio decreased the N excreted by the pig and the odorants produced in the resultant slurries. Similarly, Leek et al. (2007) reported a quadratic effect between odor emission rate and the level of dietary CP (130, 160, 190 and 210 g/kg). The authors observed that the reduction in odor rate between diets with 210 and 160 g/kg of CP decreased odorous products arising from the fermentation of protein. However, the odor rate increased as the concentration of dietary CP was reduced below 160 g/kg but remained lower than the odor rate of diets containing 190 or 210 g/kg of CP. These results suggest that factors other than the CP level of the diet, such as protein and starch fermentation, as well as DF sources used, might modify the production of odorants. In this respect, Miller and Varel (2003) reported that fermentation of starch in manure also contributes to the production of odorous compounds and conclude that achieving an optimum dietary balance between CP and starch may minimize odor emissions.

## 4. EFFECTS OF MODULATING LEVELS AND SOURCES OF DIETARY FIBER AS A STRATEGY TO INTERACT WITH PROTEIN SYNTHESIS AND FERMENTATION

Dietary fiber is associated with impaired nutrient utilization and reduced net energy values. However, a minimum of DF has to be included in the diet to maintain normal physiological functions in the GIT (Wenk, 2001).

In addition, the inclusion of DF in swine diets could improve gut health by different mechanisms as a result of its fermentation activity and responses. Moreover, the negative impacts of DF are determined by the fiber properties and may differ considerably between fiber sources. In fact, the physiological properties of NSP and their fermentability are poorly predictable from the monomeric composition and are more related to their physico-chemical characteristics like solubility, viscosity, physical structure and water binding capacity (WBC; Asp, 1996).

The physico-chemical characteristics of DF, mainly solubility and WBC, and the level of neutral detergent fiber in the diet have greater influence on the fermentation of DF. Soluble fiber sources are fermented at a faster rate, mainly at the proximal colon, whereas insoluble fiber sources have a higher WBC and greater degree of swelling (Oakenfull, 2001) and are fermented at the distal colon (Cho et al., 1997). However,Gutierrez et al. (2013) reported that ileal fermentation of DF was not affected by the level of insoluble fraction of the fiber or the amount of substrate present. In contrast, Graham et al. (1986) reported that the amount of soluble DF affect the degree of fermentation before the end of the ileum, and the fermentation of DF in the small intestine is directly related to the proportion of soluble DF. Soluble DF sources are quickly fermented in the proximal colon, and microbial biomass can itself be fermented in the distal colon, places where the microorganisms present become available as fermentable protein. Therefore, a strategy to prolong fermentation of CHO to distal colon would be important. Insoluble DF sources are slowly fermented and maintain the fermentation throughout the large intestine, reducing the occurrence of proteolytic fermentation in the large intestine (Shim et al., 2007). In this respect, Shim et al. (2007) reported that a mixture of slowly fermentable (high-polymer inulin) and rapidly fermentable (low-polymer inulin) sources produces more total gas production and also may to prolong the fermentation of CHO to distal colon. Although the fibers are primarily fermented in the large intestine, a significant amount of DF, mainly the soluble fraction, is fermented in the small intestine as well (Jørgensen et al., 1996, Jha et al., 2010 and Jha and Leterme, 2012). In this respect,Böhmer et al. (2005) found that more than 55% of dietary inulin, a highly fermentable and soluble fructan, was digested in the small intestine. The population of active bacteria becomes significant in the distal small intestine. As a consequence, the bacterial biomass accumulation occurring before the intestinal content reaches the large intestine contributes to a small but significant extent to the amino acid requirements of pigs (Torrallardona et al., 2003).

## 4.1. Changes in the gut microbiota and fermentation end products that alter the gut health

Beneficial effects of DF on intestinal functions and on the bacterial profile and fermentation are related to changes in the physicochemical characteristics of the digesta, including transit time, solubility, fermentability, viscosity or water retention capacity of the digesta. Differences between soluble and insoluble NSP have been shown to influence the digestive and fermentative processes in the growing pig. The fermentation of fiber sources by the GIT microbiota results in the generation of SCFA, and therefore, microbial fermentation in the hindgut has important implications for the gut health of the pig. Molist et al. (2009) reported an increase in the concentration of SCFA in the cecum of pigs fed diets containing 40–80 g wheat bran or 30–60 g SBP/kg of diet as compared with pigs fed a control diet without any fiber supplementation on 10–15 d post weaning. The authors concluded that the increases in the SCFA concentration with the NSP-rich diets could be associated with a higher water retention capacity of the digesta, which has been used as a predictor of the degradability of the DF (Drochner et al., 2004).

The inclusion of DF in the diet influences the composition and activity of microbiota in the GIT (Williams et al., 2001). This, in turn, may provide some protection against intestinal disorders. High or moderate amounts of DF increase the number of Lactobacilli and tend to decrease the number of some coliforms and reduce ammonia concentration in the small intestine (Kreuzer et al., 1998). Berrocoso et al. (2015) reported a decrease in *E. coli* and an increase of *lactobacillus* counts in the cecum of pigs at 43 d of age with the inclusion of SBP (50 g/kg) in diet as compared with control diet without any fiber supplementation. The authors attributed the beneficial effects of SBP inclusion to the fermentation of pectin present in SBP. Also, Molist et al. (2009) reported that diets with a higher amount of insoluble NSP (80 g/kg) or a combination of insoluble and soluble NSP (40 g/kg and 30 g/kg, respectively) promote a beneficial shift in the microbial colonization, with a higher butyric acid production in the large intestine and lower *enterobacteria* counts in the digesta. Similarly, Gerritsen et al. (2012) reported that the inclusion of insoluble NSP in the diet increases the stomach weight and the amylase activity in the intestinal brush border together with lower *E. coli* counts in the ileum and colon digesta. Diets low in soluble NSP, oligosaccharides or RS were found to reduce clinical dysentery in pigs experimentally infected with *Serpulina hydysentriae* ( Pluske et al., 1996 and Siba et al., 1996), which was confirmed later by Pluske et al. (1998). In a companion study, Pluske et al. (1998) also found that pigs fed a diet based on cooked white rice were fully protected against swine dysentery. Mateos et al. (2006) reported that the inclusion of up to 4.0% oat hulls into a low-fiber diet (5.5% NDF and 2.2% crude fiber) reduced PWD in weaning pigs from 21 to 41 d after weaning. Similarly, adding oat hulls to weaned pig diets based on extruded rice decreases protein fermentation, tends to decrease total biogenic amine concentrations, and decreases the incidence of PWD (Kim et al., 2008). This suggests that insoluble fiber sources like oat hulls could also reduce PWD, depending on the place where the protein is fermented in connection with fermentation of CHO and when there is an imbalance in the CHO:protein ratio entering the pig's large intestine.

## 4.2. Changes in the N emission in feces or urine and gases

Dietary fiber inclusion in pig diets also has been used extensively to reduce ammonia emission. The addition of NSP from soybean hulls and/or sugar beet pulp (SBP) in pig diets reduces the urinary N excretion ratio and, thereby, ammonia emission (Mroz et al., 2000 and Beccacia et al., 2015). In this respect, Beccacia et al. (2015) compared a control diet with a diet supplemented with 150 g/kg of orange pulp as a soluble fiber source and reported that the inclusion of soluble fiber led to a decrease of urinary N excretion ratio from 2.0 to 1.3. The authors noted that this reduction in the ratio could be due to an increase in the amount of N used for microbial protein synthesis in the hindgut, which is excreted in feces (organic N) instead of urine (urea N). However, other researchers reported discrepancies with respect to its effect on urinary N excretion ratio when more insoluble fiber sources were used. For example, Zervas and Zijlstra (2002)and Bindelle et al. (2009) reported that the inclusion of oat hulls in the diet did not reduce this ratio. Similar results were found in a study by Canh et al. (1997), where the increase in NSP content from 140 to 310 g/kg in the diet decreased the urinary N excretion ratio from 3.8 to 1.2 and the apparent N digestibility from 85 to 75%. Also, Kreuzer and Machmüller (1993) found that the addition of 100–220 g/kg NSP in pig diets reduce urinary N excretion by 20–28%. Jongbloed (2001) was able to show a clear relation between NSP content of the diet and the urinary N excretion ratio (Fig. 1).

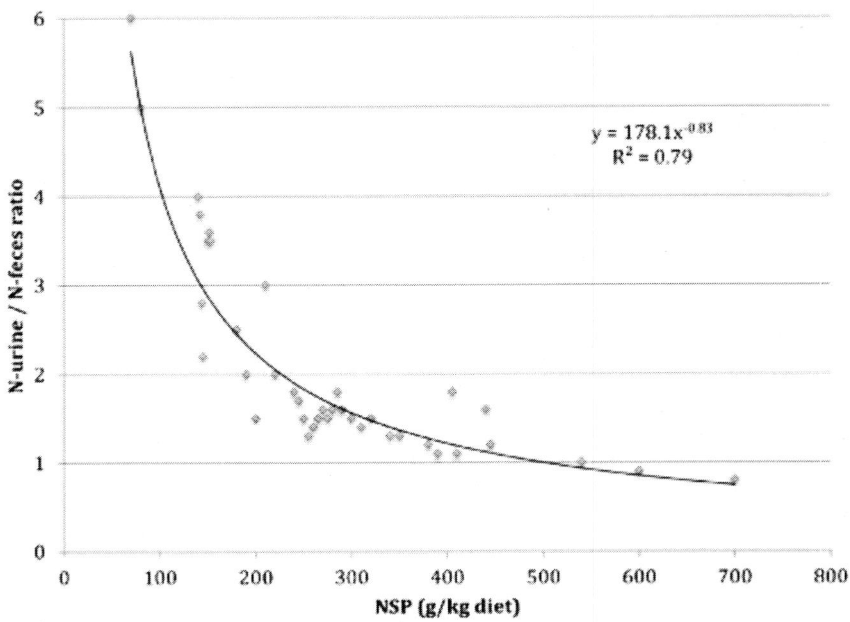


**Figure 1.** Relation between non-starch polysaccharides (NSP) content of the diet and the urine N excretion ratio (Jongbloed, 2001).

Feeds with high contents of pectin and hemicellulose, like citrus pulp and SBP, are more effective DF sources to reduce N loss in manure as compared to cellulose from rye bran and RS from cassava (Kreuzer et al., 1998). This can be ascribed to their fermentative capacity, which affects the N utilization pattern in the intestines. Sutton et al. (1999)reported that the manure of pigs fed a grain-based diet lost 2.4% of the initial N in the form of ammonia, as compared to 1.4% with SBP-based diets during a 7-d storage period. Supplementing enzymes in cereal-based diets also affects ammonia emission. Garry et al. (2007) studied the effect of diets based on barley or wheat, supplemented or not with exogenous enzymes (endo-1,3(4)-β-glucanase and endo-1,4-β-xylanase) on ammonia emission in growing-finishing pigs. They found that the addition of an enzyme to the wheat-based diet decreased the rate of ammonia emission while the enzyme's addition to the barley-based diet increased both odor and ammonia emissions. Ammonia production from manure during 10 d was higher with maize- and wheat-based diets (11.3 and 12.1% of the N intake, respectively) than with barley-based diets (6.6%). This confirms the role of the NSP in the reduction of ammonia emission.

The microbiota in the large intestine synthesizes nitrogenous cell components and secretions from simpler molecules such as ammonia, AA and peptides, using CHO as the principle energy source. With an increased supply of CHO as an energy source, any excess of indigestible protein is more likely to be incorporated into bacterial proteins than to be fermented and used as a source of energy (Houdijk et al., 1998). It is reported that 60–90% of fecal N is of bacterial origin (Rubio, 2003). Similarly, Cummings (1984)mentioned that most of the N excreted through the feces in humans and animals fed DF is considered to come from bacterial fermentation in the large intestine. The undigested protein available in the large intestine (which is of both dietary and endogenous origin) is utilized by resident bacteria as a source of N, resulting in increased bacterial proteins. The bacterial degradation of proteins starts in the stomach, as indicated by an increased formation of ammonia and, to a lesser extent, amines, phenol and p-cresol. Moreover, blood urea is the key supply of N for bacterial proliferation in the colon. In the presence of highly ureolytic bacteria in the cecum, the urea concentration gradient favors a net transfer of urea into the cecal lumen (Younes et al., 1995); thus bacteria enhance the urea transfer from blood to the large intestine (Morales et al., 2015). The ammonia generated by bacterial urease is used by bacteria for protein synthesis, which increases the amount of N present in the feces and decreases N excretion in urine, in the form of urea (Kirchgessner et al., 1994 and Younes et al., 1995). The synthesis of microbial protein causes less ammonia to be reabsorbed from the colon.

Lowering the pH of feces and manure is also an efficient means of reducing ammonia emission since ammonia is soluble under its protonated form ($NH_4^+$). Microbial fiber fermentation in the large intestine results in the acidification of the digesta due to SCFA production, resulting in the reduction of the pH of feces and manure, while the opposite is found when protein is fermented, due to ammonia production. In a series of experiments,Canh et al., 1997, Canh et al., 1998a and Canh et al., 1998c reported reduced pH of feces and manure when pigs were fed diets containing fermentable fiber sources, apart from the shift in N excretion pathway. For each increase of 100 g of non-starch polysaccharides (NSP) in pig diets, the pH of the slurry decreased by 0.12 units and ammonia emission was reduced by 5.4%. The decrease in pH was partly attributed to the decrease in the $NH_4$-N content of the slurry as well as to increased SCFA concentration of the slurry following CHO fermentation (Canh et al., 1998a). Similarly, lowering the pH of urine can lower the pH of manure, thus reducing ammonia release from the storage. The urine pH can be altered by changing the electrolyte balance of the diet (Patience et al., 1987).

### 4.3. Changes in odorous compounds

Changes in DF content are also purported to affect odor emissions from manure, although the effect is more complicated than is the case for protein reduction. The presence of NSP influences the bacterial dynamics in the hindgut and directly impacts the production of odorous compounds such as SCFA, amines, and sulfides. In this respect,Canh et al. (1997) reported that an increases of fermentable CHO in diets increases manure SCFA concentrations. This increase may impact manure odor concentration, because SCFA are important odorous compounds in manure storage. De Camp et al. (2001) compared a control diet with diets supplemented with 100 g/kg soybean hulls as fiber sources and reported a 32% increase of total SCFA concentration in 6-week-stored manure and an 11% reduction in odor concentration. However, Gralapp et al. (2002)reported no difference in odor concentration when 100 g/kg distillers dried grain was added to the diets of finishing pigs. Finally, the pattern of odor emissions could be due to a number of reasons; for instance, limiting protein fermentation and increasing CHO fermentation in the large intestine can reduce the production of some of the most offensive-smelling compounds emanating from pig production. In this respect, both protein and CHO fermentation contribute to the production of acetic acid; however, the BCFAs, i.e. iso-butyric and iso-valeric acid, are produced exclusively from protein fermentation and have a more offensive odor than the straight-chain SCFAs (Mackie et al., 1998).

## 5. CONCLUSION

The reduction in CP content in the diet (less than 190 g/kg), together with the inclusion of moderately fermentable fiber sources will reduce the production of harmful microbial metabolites in the large intestine as well as the incidence of intestinal disorders in pigs. There is high bacterial N uptake when pigs are fed with fermentable fiber sources, suggesting that, in presence of higher levels of fiber substrate for fermentation, the resident microbiota in the large intestine retain more N for their own growth. There is, thus, increased bacterial protein mass and also a shift in N excretion pathway from urine to feces. The inclusion of soluble fiber leads to a decrease of urinary N excretion. Finally, dietary CP and DF manipulation is an effective tool to reduce nitrogenous gases from piggeries. Nitrogen excretion, ammonia emission from pigs, and manure concentrations of odorous compounds can be reduced substantially by strategically decreasing dietary protein and increasing fermentable fiber in pig diets, although the effect of changes in content of DF is relatively less complex than is the case for protein reduction. Therefore, the inclusion of DF and reduction in dietary protein content can be used as a nutritional strategy to optimize intestinal health of early-weaned pigs.

## ACKNOWLEDGEMENT

Postdoctoral fellow Dr Julio F. D. Berrocoso was partly supported from the USDA, Agricultural Research Service, Collaborative Research Fund (Agreement no. 58-5320-3-022).

## REFERENCES

1. Aarnink, A.J.A., Hoeksma, P., Van Ouwerkerk, E.N.J., 1993. Factors affecting ammonium concentration in slurry from fattening pigs. In: Verstegen, M.W.A., den Hartog, L.A., van Kempen, G.J.M., Metz, J.H.M. (Eds.), Nitrogen Flow in Pig Production and Environmental Consequences, Pudoc-DLO. EAAP-Publication, Wageningen, The Netherlands, pp. 413–420.

2. Aarnink, A.J.A., Swiestra, D., van den Berg, A.J., Speelman, L., 1997. Effect of type of slatted floor and degree of fouling of solid floor on ammonia emission rates from fattening piggeries. J. Agr. Eng. Res. 66, 93–102.

3. Aarnink, A.J.A., Verstegen, M.W.A., 2007. Nutrition, key factor to reduce environmental load from pig production. Livest. Sci. 109, 194–203.

4. Asp, N.G., 1996. Dietary carbohydrates: classification by chemistry and physiology. Food Chem. 57, 9–14.

5. Aumiller, T., Mosenthin, R., Weiss, E., 2015. Potential of cereal grains and grain legumes in modulating pigs' intestinal microbiota – A review. Livest. Sci. 172, 16–32.

6. Ball, R.O., Möhn, S., 2003. Feeding strategies to reduce greenhouse gas emissions from pigs. Advances Pork Prod. 14, 301–311.

7. Beccacia, A., Calvet, S., Cerisuelo, A., Ferrer, P., García-Rebollar, P., De Blas, C., 2015. Effects of nutrition on digestion efficiency and gaseous emissions from slurry in growing-finishing pigs. I. Influence of the inclusion of two levels of orange pulp and carob meal in iso fibrous diets. Anim. Feed Sci. Technol. 208, 158–169.

8. Berrocoso, J.D., Menoyo, D., Guzmán, P., Saldana, ˜ B., Cámara, L., Mateos, G.G., 2015. Effects of fiber inclusion on growth performance and nutrient digestibility of piglets reared under optimal or poor hygienic conditions. J. Anim. Sci. 93, 3919–3931.

9. Bikker, P., Dirkzwager, A., Fledderus, J., Trevisi, P., Huerou-Luron, I.L., Lalles, J.P., Awati, A., 2006. The effect of dietary protein and fermentable carbohydrates levels on growth performance and intestinal characteristics in newly weaned piglets. J. Anim. Sci. 84, 3337–3345.

10. Bindelle, J., Buldgen, A., Delacollette, M., Wavreille, J., Agneessens, R., Destain, J.P., Leterme, P., 2009. Influence of source and concentrations of dietary fiber on in vivo nitrogen excretion pathways in pigs reflected by in vitro fermentation and N incorporation by fecal bacteria. J. Anim. Sci. 87, 583–593.

11. Blachier, F., Mariotti, F., Huneau, J.F., Tome, D., 2007. Effects of amino acid-derived luminal metabolites on the colonic epithelium and physiopathological consequences. Amino Acids 33, 547–562.

12. Böhmer, B.M., Branner, G.R., Roth-Maier, D.A., 2005. Precaecal and faecal digestibility of inulin (DP 10-12) or an inulin/Enterococcus faecium mix and effects on nutrient digestibility and microbial gut flora. J. Anim. Physiol. Anim. Nutr. 89, 388–396.

13. Canh, T.T., Verstegen, M.W.A., Aarnink, A.J.A., Schrama, J.W., 1997. Influence of dietary factors on nitrogen partitioning and composition of urine and feces of fattening pigs. J. Anim. Sci. 75, 700–706.

14. Canh, T.T., Aarnink, A.J., Verstegen, M.W., Schrama, J.W., 1998a. Influence of dietary factors on the pH and ammonia emission of slurry from growing-finishing pigs. J. Anim. Sci. 76, 1123–1130.

15. Canh, T.T., Aarnink, A.J.A., Schutte, J.B., Sutton, A., Langhout, D.J., Verstegen, M.W.A., 1998b. Dietary protein affects nitrogen excretion and ammonia emission from slurry of growing-finishing pigs. Livest. Prod. Sci. 56, 181–191.

16. Canh, T.T., Sutton, A.L., Aarnink, A.J., Verstegen, M.W., Schrama, J.W., Bakker, G.C., 1998c. Dietary carbohydrates alter the fecal composition and pH and the ammonia emission from slurry of growing pigs. J. Anim. Sci. 76, 1887–1895.

17. Cho, S., DeVries, J.W., Prosky, L., 1997. Dietary Fiber Analysis and Applications. AOAC Intl., Gaithersburg, MD, USA.

18. Cone, J.W., Jongbloed, A.W., Gelder, A.H.V., Lange, L.D., 2005. Estimation of protein fermentation in the large intestine of pigs using a gas production technique. Anim. Feed Sci. Technol. 123-124, 463–472.

19. Cummings, J.H., 1984. Microbial digestion of complex carbohydrates in man. Proc. Nutr. Soc. 43, 35–44.

20. Davila, A.M., Blachier, F., Gotteland, M., Andriamihaja, M., Benetti, P.H., Sanz, Y., Tome, D., 2013. Intestinal luminal nitrogen metabolism: role of the gut microbiota and consequences for the host. Pharm. Res. 68, 95–107.

21. De Camp, S.A., Hill, B.E., Hankins, S.L., Bundy, D.C., Powers, W.J., 2001. Effects of soybean hulls in commercial diet on pig performance, manure composition, and selected air quality parameters in swine facilities. J. Anim. Sci. 79 (Suppl. 1), S252.

22. Drochner, W., Kerler, A., Zacharias, B., 2004. Pectin in pig nutrition, a comparative review. J. Anim. Physiol. Anim. Nutr. 88, 367–380.

23. Garry, B.P., Fogarty, M., Curran, T.P., O'Connell, M.J., O'Doherty, J.V., 2007. The effect of cereal type and enzyme addition on pig performance, intestinal microflora, and ammonia and odour emissions. Animal 1, 751–757.

24. Gerritsen, R., van der Aar, P., Molist, F., 2012. Insoluble nonstarch polysaccharides in diets for weaned piglets. J. Anim. Sci. 90, 318–320.

25. Graham, H., Hesselman, K., Aman, P., 1986. The influence of wheat bran and sugar-beet pulp on the digestibility of dietary components in a cereal-based pig diet. J. Nutr. 116, 242–251.

26. Gralapp, A.K., Powers, W.J., Faust, M.A., Bundy, D.S., 2002. Effects of dietary ingredients on manure characteristics and odorous emissions from swine. J. Anim. Sci. 80, 1512–1519.

27. Gutierrez, N.A., Kerr, B.J., Patience, J.F., 2013. Effect of insoluble-low fermentable fiber from corn-ethanol distillation origin on energy, fiber, and amino acid digestibility, hindgut degradability of fiber, and growth performance of pigs. J. Anim. Sci. 91, 5314–5325.

28. Heo, J.M., Kim, J.C., Hansen, C.F., Mullan, B.P., Hampson, D.J., Pluske, J.R., 2009. Feeding a diet with decreased protein content reduces indices of protein fermentation and the incidence of postweaning diarrhea in weaned pigs challenged with an enterotoxigenic strain of Escherichia coli. J. Anim. Sci. 87, 2833–2843.

29. Heo, J.M., Kim, J.C., Hansen, C.F., Mullan, B.P., Hampson, D.J., Pluske, J.R., 2010. Feeding a diet with a decreased protein content reduces both nitrogen content in the gastrointestinal tract and post-weaning diarrhoea, but does not affect apparent nitrogen digestibility in weaner pigs challenged with an enterotoxigenic strain of Escherichia coli. Anim. Feed Sci. Technol. 160, 48–159.

30. Hobbs, P.J., Pain, B.F., Kay, R.M., Lee, P.A., 1996. Reduction of odorous compounds in fresh pig slurry by dietary control of crude protein. J. Sci. Food Agric. 71, 508–514.

31. Houdijk, J.G.M., Bosch, M.W., Verstegen, M.W.A., Berenpas, H.J., 1998. Effects of dietary oligosaccharides on the growth performance and faecal characteristics of young growing pigs. Anim. Feed Sci. Technol. 71, 35–48.

32. Htoo, J.K., Araiza, B.A., Sauer, W.C., Rademacher, M., Zhang, Y., Cervantes, M., Zijlstra, R.T., 2007. Effect of dietary protein content on ileal amino acid digestibility, growth performance, and formation of microbial metabolites in ileal and cecal digesta of early-weaned pigs. J. Anim. Sci. 85, 3303–3312.

33. Jha, R., Leterme, P., 2012. Feed ingredients differing in fermentable fibre and indigestible protein content affect fermentation metabolites and faecal nitrogen excretion in growing pigs. Animal 6, 603–611.

34. Jha, R., Berrocoso, J.D., 2015. Review: Dietary fiber utilization and its effects on physiological functions and gut health of swine. Animal 9, 1441–1452.

35. Jha, R., Rossnagel, B., Pieper, R., Van Kessel, A., Leterme, P., 2010. Barley and oat cultivars with diverse carbohydrate composition alter ileal and total tract nutrient digestibility and fermentation metabolites in weaned piglets. Animal 4, 724–731.

36. Jha, R., Htoo, J.K., Young, M.G., Beltranena, E., Zijlstra, R.T., 2013. Effect of increasing co-product inclusion and reducing dietary protein on growth performance, carcass characteristics, and jowl fatty acid profile of grower-finisher pigs. J. Anim. Sci. 91, 2178–2219.

37. Jensen, M.T., Cox, R.P., Jensen, B.B., 1995. 3- Methylindole (skatole) and indole production by mixed populations of pig fecal bacteria. Appl. Environ. Microbiol. 61, 3180–3184.

38. Johnson, L.R., 2001. Digestion and absorption. In: Johnson, L.R. (Ed.), Gastrointestinal Physiology. Mosby, St. Louis, MO, USA, p. p120. Jongbloed, A.W., 2008. Environmental pollution control in pigs by using nutrition tools. Rev. Bras. Zootec. 37, 215–229.

39. Jongbloed, A.W., 2001. Hebben voermaatregelen ter verlaging van de ammoniakemissie een negatief effect op de vertering en benutting van nutriënten? (Do feeding measures for lowering ammonia emission have a negative effect on the digestibility and utilisation of nutrients?). In: Jongbloed, A.W., Blok, M.C. (Eds.), Veevoeding en Ammoniakemissie uit Varkensstallen. Samenvattingen van de themamiddag, 2174. Productschap Diervoeder, The Hague, The Netherlands, pp. 11–23.

40. Jorgensen, H., Zhao, X.Q., Eggum, B.O., 1996. The influence of dietary fibre and environmental temperature on the development of the gastrointestinal tract, digestibility, degree of fermentation in the hind-gut and energy metabolism in pigs. Br. J. Nutr. 75, 365–378.

41. Kerr, B.J., 1995. Nutritional strategies for waste reduction-management: nitrogen. In: Longenecker, J.B., Spears, J.W. (Eds.),

New Horizons in Animal Nutrition and Health. The Inst. Nutr. Univ. North Carolina, Chapel Hill, Raleigh, NC, USA, pp. 47–68.

42. Kim, J.C., Mullan, B.P., Hampson, D.J., Pluske, J.R., 2008. Addition of oat hulls to an extruded rice-based diet for weaner pigs ameliorates the incidence of diarrhoea and reduces indices of protein fermentation in the gastrointestinal tract. Br. J. Nutr. 99, 1217–1225.

43. Kirchgessner, M., Kreuzer, M., Machmuller, A., Roth-Maier, D.A., 1994. Evidence for a high efficiency of bacterial protein synthesis in the digestive tract of adult sows fed supplements of fibrous feedstuffs. Anim. Feed Sci. Technol. 46, 293–306.

44. Kreuzer, M., Machmüller, A., 1993. Reduction of gaseous nitrogen emission from pig manure by increasing the level of bacterially fermentable substrates in the ration. In: Verstegen, M.W.A., den Hartog, L.A., van Kempen, G.J.M., Metz, J.H.M. (Eds.), Nitrogen Flow in Pig Production and Environmental Consequences. EAAP-Publication, Pudoc-DLO, Wageningen, The Netherlands, pp. 151–156.

45. Kreuzer, M., Machmüller, A., Gerdemann, M.M., Hanneken, H., Wittmann, M., 1998. Reduction of gaseous nitrogen loss from pig manure using feeds rich in easily-fermentable non-starch polysaccharides. Anim. Feed Sci. Technol. 73, 1–19.

46. Kröger, S., Pieper, R., Schwelberger, H.G., Wang, J., Villodre Tudela, C., Aschenbach, J.R., Van Kessel, A.G., Zentek, J., 2013. Diets high in heat treated soybean meal reduce the histamine-induced epithelial response in the colon of weaned piglets and increase epithelial catabolism of histamine. PLOS One 8, e80612.

47. Le, P.D., Aarnink, A.J.A., Jongbloed, A.W., Peet-Schwering, C.M.C., van, der., Ogink, N.W.M., Verstegen, M.W.A., 2008. Interactive effects of dietary crude protein and fermentable carbohydrate levels on odour from pig manure. Livest. Sci. 114, 48–61.

48. Leek, A.B.G., Callan, J.J., Reilly, P., Beattie, V.E., O'Doherty, J.V., 2007. Apparent component digestibility and manure ammonia emission in finishing pigs fed diets based on barley, maize or wheat prepared without or with exogenous non-starch polysaccharide enzymes. Anim. Feed Sci. Technol. 135, 86–99.

49. Lenis, N.P., 1989. Lower nitrogen excretion in pig husbandry by feeding: Current and future possibilities. Neth. J. Agri. Sci. 37, 61–70.

50. Macfarlane, G.T., Gibson, G.D., 1995. Microbiological aspects of the production of short-chain fatty acids in the large bowel. In: Cumming, J.H., Rombeau, J.L., Sakota, T. (Eds.), Physiological and Clinical Aspects of Short-Chain Fatty Acids. Cambridge University Press, Cambridge, UK, pp. 87–105.

51. Macfarlane, S., Macfarlane, G.T., 2003. Regulation of short-chain fatty acid production. Proc. Nutr. Soc. 62, 67–72.

52. Macfarlane, G.T., Gibson, G.R., Beatty, E., Cummings, J.H., 1992. Estimation of short-chain fatty acid production from protein by human intestinal bacteria based on branched-chain fatty acid measurements. FEMS Microbiol. Lett. 101, 81–88.

53. Mackie, R.I., Stroot, P.G., Varel, V.H., 1998. Biochemical identification and biological origin of key odor components in livestock waste. J. Anim. Sci. 76, 1331–1342.

54. Mateos, G.G., Martin, F., Latorre, M.A., Vicente, B., Lazaro, R., 2006. Inclusion of oat hulls in diets for young pigs based on cooked maize or cooked rice. Anim. Sci. 82, 57–63.

55. Miller, D.N., Varel, V.H., 2003. Swine manure composition affects the biochemical origins, composition, and accumulation of odorous compounds. J. Anim. Sci. 81, 2131–2138.

56. Möhn, S., Susenbeth, A., 1995. Influence of dietary protein content on efficiency of energy utilisation in growing pigs. Arch. Anim. Nutr. 47, 361–372.

57. Molist, F., Gómez de Segura, A., Gasa, J., Hermes, R.G., Manzanilla, E.G., Anguita, M., Pérez, J.F., 2009. Effects of the insoluble and soluble dietary fibre on the physicochemical properties of digesta and the microbial activity in early weaned piglets. Anim. Feed Sci. Technol. 149, 346–353.

58. Montagne, L., Pluske, J.R., Hampson, D.J., 2003. A review of interactions between dietary fibre and the intestinal mucosa, and their consequences on digestive health in young non-ruminant animals. Anim. Feed Sci. Technol. 108, 95–117.

59. Morales, A., Buenabad, L., Castillo, G., Arce, N., Araiza, B.A., Htoo, J.K., Cervantes, M., 2015. Low-protein amino acid–supplemented diets for growing pigs: Effect on expression of amino acid transporters, serum concentration, performance, and carcass composition. J. Anim. Sci. 93, 2154–2164.

60. Mroz, Z., Moeser, A.J., Vreman, K., van Diepen, J.T., van Kempen, T., Canh, T.T., Jongbloed, A.W., 2000. Effects of dietary carbohydrates and buffering capacity on nutrient digestibility and manure characteristics in finishing pigs. J. Anim. Sci. 78, 3096–3106.

61. Nollet, H., Deprez, P., Van Driessche, E., Muylle, E., 1999. Protection of just weaned pigs against infection with F18 + Escherichia coli by non-immune plasma powder. Vet. Microbiol. 65, 37–45.

62. Nousiainen, J.T., 1991. Comparative observations on selected probiotics and olaquindox as feed additives for piglets around weaning. 2. Effect on villus length and crypt depth in the jejunum, ileum, caecum and colon. J. Anim. Physiol. Anim. Nutr. 66, 224–230.

63. Oakenfull, D., 2001. Dietary fiber in human nutrition. In: Spiller, G.A. (Ed.), Physical Chemistry of Dietary Fiber. CRC Press, Boca Raton, FL, USA, pp. 33–47.

64. Patience, J.F., Austic, R.E., Boyd, R.D., 1987. Effect of dietary electrolyte balance on growth and acid-base status in swine. J. Anim. Sci. 64, 457–466.

65. Pieper, R., Jha, R., Rossnagel, B., Van Kessel, A.G., Souffrant, W.B., Leterme, P., 2008. Effect of barley and oat cultivars with different carbohydrate compositions on the intestinal bacterial communities in weaned piglets. FEMS Microbiol. Ecol. 66, 556–566.

66. Pluske, J.R., Siba, P.M., Pethick, D.W., Durmic, Z., Mullan, B.P., Hampson, D.J., 1996. The incidence of swine dysentery in pigs can be reduced by feeding diets that limit the amount of fermentable substrate entering the large intestine. J. Nutr. 126, 2920–2933.

67. Pluske, J.R., Durmic, Z., Pethick, D.W., Mullan, B.P., Hampson, D.J., 1998. Confirmation of the role of rapidly fermentable

carbohydrates in the expression of swine dysentery in pigs after experimental infection. J. Nutr. 128, 1737–1744.

68. Rubio, L.A., 2003. Carbohydrates digestibility and faecal nitrogen excretion in rats fed raw or germinated faba bean (Vicia faba)- and chickpea (Cicer arietinum)-based diets. Br. J. Nutr. 90, 301–309.

69. Siba, P.M., Pethick, D.W., Hampson, D.J., 1996. Pigs experimentally infected with Serpulina hyodysenteriae can be protected from developing swine dysentery by feeding them a highly digestible diet. Epidemiol. Infect. 116, 207–216.

70. Shim, S.B., Verdonk, J.M.A.J., Pellikaan, W.F., Verstegen, M.W.A., 2007. Differences in microbial activities of faeces from weaned and unweaned pigs in relation to in vitro fermentation of different sources of inulin-type oligofructose and pig feed ingredients. Asian-Aust. J. Anim. Sci. 20, 1444–1452.

71. Sutton, A.L., Kephart, K.B., Verstegen, M.W.A., Canh, T.T., Hobbs, P.J., 1999. Potential for reduction of odorous compounds swine manure through diet modification. J. Anim. Sci. 77, 430–439.

72. Torrallardona, D., Harris, C.I., Fuller, M.F., 2003. Pigs'gastrointestinal microflora provide them with essential amino acids. J. Nutr. 133, 1127–1131.

73. Tso, P., Crissinger, K., 2000. Overview of digestion and absorption. In: Stipanuk, M.H., Caudill, M.A. (Eds.), Biochemical and Physiological Aspects of Human Nutrition. Saunders, PA, USA, pp. 75–106.

74. Urbain, B., Gustin, P., Prouvost, J.F., Ansay, M., 1994. Quantitative assessment of aerial ammonia toxicity to the nasal mucosa by use of the nasal lavage method in pigs. Am. J. Vet. Res. 55, 1335–1340.

75. Williams, B.A., Verstegen, M.W.A., Tamminga, S., 2001. Fermentation in the large intestine of single-stomached animals and its relationship to animal health. Nutr. Res. Rev. 14, 207–227.

76. Wellock, I.J., Fortomaris, P.D., Houdijk, J.G.M., Wiseman, J., Kyriazakis, I., 2008. The consequences of non-starch polysaccharide solubility and inclusion level on the health and performance of weaned pigs challenged with enterotoxigenic Escherichia coli. Br. J. Nutr. 99, 520–530.

77. Wenk, C., 2001. The role of dietary fibre in the digestive physiology of the pig. Anim. Feed Sci. Technol. 90, 21–33.

78. Younes, H., Garleb, K., Behr, S., Remesy, C., Demigne, C., 1995. Fermentable fibers or oligosaccharides reduce urinary nitrogen excretion by increasing urea disposal in the rat cecum. J. Nutr. 125, 1010–1016.

79. Zervas, S., Zijlstra, R.T., 2002. Effects of dietary protein and fermentable fiber on nitrogen excretion patterns and plasma urea in grower pigs. J. Anim. Sci. 80, 3247–3256.

80. Zhang, Y., Tanaka, A., Dosman, J.A., Senthilselvan, A., Barber, E.M., Kirychuk, S.P., Holfeld, L.E., Hurst, T.S., 1998. Acute respiratory responses of human subjects to air quality in a swine building. J. Agr. Eng. Res. 70, 367–373.

# CHAPTER 11

# Comparative effects of dietary carbohydrases without or with protease on the ileal digestibility of energy and amino acids and $AME_n$ in young broilers

**L.F. Romero[a], C.M. Parsons[b], P.L. Utterback[b], P.W. Plumstead[a], V. Ravindran[c]**

[a] Danisco Animal Nutrition – DuPont Industrial Biosciences, Marlborough, SN8 1XN, UK

[b] Department of Animal Science, University of Illinois, Urbana, IL 61801, USA

[c] Institute of Food, Nutrition and Human Health, Massey University, Palmerston North 4442, New Zealand

## ABSTRACT

Four digestibility trials using broilers from 12 to 21 d or 7 to 21 d of age were conducted to evaluate the nutrient digestibility responses of two enzyme combinations when supplemented to maize–soybean meal (SBM) based diets without (trials 1 and 3) or with (trials 2 and 4) 70 or 100 g/kg maize dried distillers grain with solubles (DDGS). Apparent ileal digestible energy (AIDE), apparent ileal digestibility (AID) of amino acids (AA), total tract apparent retention (TTAR) of nitrogen (N) and N-corrected apparent metabolisable energy ($AME_n$) were assessed.

In each trial, a control diet and the control diet supplemented with an enzyme complex containing xylanase and amylase (XA) or one containing protease, xylanase and amylase (PXA) were tested. Trials 1 and 2 had six replicate cages per treatment, with six birds per cage; and trials 3 and 4 had eight replicate cages per treatment, with five birds per cage. Data were analysed by ANOVA for each trial, as well as a combined dataset, which model included the main effects of maize-DDGS inclusion, dietary enzyme and trial site and all two-way interactions. Across trials, both XA (13.61 MJ/kg) and PXA (13.70 MJ/kg) increased the AIDE ($P<0.05$) compared to the control diets (13.29 MJ/kg). XA increased ($P<0.05$) $AME_n$ (12.89 MJ/kg) compared to the control diets (12.61 MJ/kg), and PXA further increased it (13.00 MJ/kg) compared to XA. Supplementation with XA had no effect on the AID of AA, whereas PXA increased the AID of nitrogen and all AA ($P<0.05$) except methionine. The AA with the greatest response to PXA were cysteine (+5.4%), threonine (+4.4%), glycine (+3.6%) and valine (+3.3%). The least responsive AA to PXA inclusion were methionine (+1.0), glutamine (+2.0), lysine (+1.9%) and arginine (+2.2%). Irrespective of the AA, PXA increased AID of AA as a proportion of the ileal undigested fraction of AA in the control diets by 12–13%. Supplemental xylanase and amylase increased AIDE and $AME_n$ and the addition of protease on top of carbohydrases further increased AID of AA and $AME_n$ in young broilers fed maize–SBM diets without or with maize-DDGS.

## ABBREVIATIONS

AA - amino acids

AME - apparent metabolisable energy

$AME_n$ - nitrogen corrected apparent metabolisable energy

AID - apparent ileal digestibility

AIDE - apparent ileal digestible energy

DDGS - dried distillers grain with solubles

N - nitrogen

NSPs - non-starch polysaccharides

PXA - protease, xylanase and amylase

TTAR - total tract apparent retention

XA - xylanase and amylase

## KEYWORDS

Amino acid; Broiler; Carbohydrase; Digestibility; Maize-DDGS; Protease

## 1. INTRODUCTION

Carbohydrase enzymes such as xylanases, amylases and glucanases are being adopted by the poultry industry worldwide. In wheat-based diets, these enzymes are used to ameliorate the adverse effects of non-starch polysaccharides (NSPs) on digesta viscosity, nutrient digestibility and bird performance (Choct, 2006). Exogenous carbohydrases are now increasingly used in maize-based diets for broilers (Cowieson, 2010) despite low concentrations of soluble NSP in these diets. Economic benefits from the use of these enzymes are obtained through a reduction of feed cost that occurs when improvements on apparent metabolisable energy (AME) and, to a lesser extent, amino acids (AA) digestibility are accounted for in diet formulations. Possible mechanisms of action of carbohydrases in poultry diets include: improved access of endogenous enzymes to cell contents due to hydrolysis of cell wall arabinoxylans (Cowieson, 2005), augmentation of digestive enzymes in young animals, particularly of amylase (Ritz et al., 1995 and Gracia et al., 2003), modulation of intestinal microflora (Fernandez et al., 2000) and reduction of endogenous AA losses, particularly through changes on pancreatic amylase (Jiang et al., 2008) and mucin secretion (Cowieson and Bedford, 2009). In general, improvements of AME due to exogenous xylanase and amylase appear to be a combination of two mechanisms, namely, increased digestion and absorption of the undigested portion of starch and fat of the diet and down-regulation of various digestive secretions.

Exogenous proteases of microbial origin are being increasingly used in broiler diets (Adeola and Cowieson, 2011). The economic benefit of the use of exogenous proteases is through improvements in the digestibility of dietary AA. The primary mechanism for this increment appears to be the augmentation of dietary protein hydrolysis and increased protein solubility (Caine et al., 1998). Conclusive evidence is not available supporting a positive or negative change in endogenous secretions caused by exogenous proteases. However, increments of AME caused by exogenous proteases in conjunction with xylanases and amylases are suggestive of the ability of exogenous proteases to disrupt protein–starch interactions in cereal grains (McAllister et al., 1993 and Belles et al., 2000). Nonetheless, the additional effect of protease on top of carbohydrases on energy metabolisability and apparent ileal digestibility (AID) of AA has not been fully assessed. That is particularly relevant for modern broiler diets containing non-traditional protein sources such as maize dried distillers grain with solubles (DDGS), where the effects of proteases and carbohydrases may differ in importance compared to traditional maize–soybean meal (SBM) diets. The current study aimed to assess the effect of an exogenous bacterial serine protease in combination with a xylanase and an amylase *versus* xylanase and amylase only in broilers fed either a maize–SBM, or a maize–SBM-DDGS based diet, in terms of apparent ileal digestible energy (AIDE), AID of AA, total tract apparent retention (TTAR) of nitrogen (N) and N-corrected AME ($AME_n$). The results reported herein were from four independent trials with similar design performed at two research sites.

# 2. MATERIALS AND METHODS

## 2.1. Exogenous enzymes

Two enzyme products were used: a prototype enzyme preparation containing endo-1,4-beta-xylanase (EC 3.2.1.8) and alpha-amylase (EC 3.2.1.1); and a commercial enzyme preparation (Avizyme 1502; Danisco Animal Nutrition DuPont Industrial Biosciences, Marlborough, UK) containing endo-1,4-beta-xylanase (EC 3.2.1.8), alpha-amylase (EC 3.2.1.1) and an alkaline serine protease (EC 3.4.21.62). The xylanase originated from*Trichoderma reesei*, the amylase from *Bacillus amyloliquifaciens* and the protease from*Bacillus subtilis*.

## 2.2. Experimental design

Four digestibility trials were conducted to evaluate AIDE, AID of AA and $AME_n$ of broilers fed maize–SBM diets without (two trials) or with (two trials) maize-DDGS inclusion, supplemented with an enzyme complex containing xylanase and amylase (XA), or one containing protease, xylanase and amylase (PXA), as compared to a control diet. In each trial, three dietary treatments (control, XA and PXA) were evaluated in diets for broilers from hatch to 21 d of age. Trials 1 and 2, conducted at Massey University (Palmerston North, New Zealand) consisted of six replicates per treatment, with six birds per replicate cage. Trials 3 and 4, conducted at the University of Illinois (Urbana, USA), consisted of eight replicates per treatment, with five birds per replicate cage.

## 2.3. Experimental diets

Ingredients and calculated diet composition are presented in Table 1. The diets were based on maize and SBM, with slight variations in ingredient composition between the trial sites. Diets were provided in mash form. In each research site, one trial was performed without and one with maize-DDGS inclusion. Maize-DDGS was included at 100 g/kg in diets of trial 2 and 70 g/kg in trial 4. Calculated AME values were constant within site. Titanium dioxide (3.0 g/kg) was added to all diets as an indigestible marker. Diets were manufactured in one batch for each trial and then subdivided in three experimental diets, two of them containing enzymes. Concentrates of the test enzymes were sprayed into a wheat carrier and added to the respective diets in dry form at 0.5 g/kg, after being premixed with 5 g of maize/kg from the diets.

Enzyme activities in feed samples (200 g) were measured at the Danisco Innovation Laboratories (Brabrand, Denmark; Table 2) in duplicate. One xylanase unit was defined as the amount of enzyme that releases 0.48 μmol of reducing sugar as xylose from wheat arabino xylan per minute at pH 4.2 and 50 °C. Azurine cross linked arabinoxylan isolated from wheat (Megazyme International Ireland Ltd., Bray, Ireland) was used as substrate. Sample extracts (100 μl) were incubated at 50 °C for 60 min, mixed with a vortex and centrifuged at 960×*g* for 10 min, after which absorbance of the supernatant was measured with a spectrophotometer at 590 nm against a blank sample and units were calculated in reference to a calibration curve.

**Table 1.** Composition and analysis of basal diets of four trials using maize–soybean meal based diets without (trials 1 and 3) or with (trials 2–4) 70 or 100 g/kg maize-DDGS.

| | Trial 1 | Trial 2 | Trial 3 | Trial 4 |
|---|---|---|---|---|
| Ingredients (g/kg) | | | | |
| Maize | 606.0 | 548.4 | 577.0 | 539.7 |
| Soybean meal (480 g/kg) | 337.9 | 289.1 | 372.1 | 339.0 |
| Maize-DDGS | 0.0 | 100.0 | 0.0 | 70.0 |
| Soy oil | 15.0 | 22.1 | 12.4 | 13.1 |
| Dicalcium phosphate | 13.8 | 12.1 | 13.5 | 12.5 |
| Limestone | 11.7 | 12.9 | 11.7 | 12.1 |
| Sodium bicarbonate | 2.7 | 1.6 | 0.0 | 0.0 |
| Salt | 2.0 | 2.3 | 3.5 | 3.5 |
| L-Lysine HCl (780 g/kg) | 1.8 | 2.7 | 0.5 | 1.0 |
| DL-Methionine (990 g/kg) | 2.6 | 2.2 | 2.3 | 2.1 |
| L-Threonine, (980 g/kg) | 0.0 | 0.1 | 0.0 | 0.0 |
| Broiler premix[a,b] | 3.0 | 3.0 | 3.5 | 3.5 |
| Titanium dioxide | 3.0 | 3.0 | 3.0 | 3.0 |
| Starch or enzyme | 0.5 | 0.5 | 0.5 | 0.5 |
| Determined composition (g/kg) | | | | |
| Dry matter | 894 | 890 | 886 | 892 |
| $AME_n$ (MJ/kg) | 11.3 | 11.3 | 11.0 | 11.3 |
| Crude protein | 198 | 200 | 195 | 201 |
| Lysine | 12.7 | 11.6 | 11.7 | 11.7 |
| Methionine | 6.3 | 6.4 | 4.9 | 4.8 |
| Methionine + cysteine | 9.0 | 9.6 | 8.0 | 7.9 |
| Calculated nutrient composition (g/kg) | | | | |
| AME (MJ/kg) | 12.5 | 12.5 | 12.4 | 12.4 |
| Crude protein | 217 | 217 | 230 | 229 |
| Calcium | 8.5 | 8.5 | 8.5 | 8.5 |
| Available phosphorus | 3.8 | 3.8 | 3.8 | 3.8 |
| Lysine | 13.0 | 13.0 | 13.0 | 13.0 |
| Methionine | 6.0 | 5.9 | 5.8 | 5.7 |
| Methionine + cysteine | 9.5 | 9.5 | 9.5 | 9.5 |

[a] Trials 1 and 2. The premix supplied per kilogram of diet: biotin, 0.2 mg; D-Ca-pantothenate, 12.8 mg; cholecalciferol, 60 µg; cyanocobalamin, 0.017 mg; folic acid, 5.2 mg; menadione, 4 mg; niacin,

35 mg; pyridoxine, 10 mg; trans-retinol, 3.33 mg; riboflavin, 12 mg; thiamine, 3.0 mg; DL-α-tocopheryl acetate, 60 mg; choline chloride, 638 mg; Co, 0.3 mg; Cu, 3 mg; Fe, 25 mg; I, 1 mg; Mn, 125 mg; Mo, 0.5 mg; Se, 200 μg; Zn, 60 mg; and antioxidant, 100 mg.

[b] Trials 3 and 4. The premix supplied per kilogram of diet; D-Ca-pantothenate, 10 mg; cholecalciferol, 25 μg; cyanocobalamin, 0.01 mg; menadione sodium bisulphite, 2.33 mg; retinyl acetate, 4400 IU; riboflavin, 4.41 mg; DL-α-tocopheryl acetate, 11 IU; niacin, 22 mg; Cu, 5 mg; Fe, 75 mg; Mn, 75 mg; Se, 0.1 mg; Zn, 75 mg; I, 0.75 mg.

One unit of amylase was defined as the amount of the enzyme catalysing the hydrolysis of 1 μmol glucosidic linkage per minute. Amylase activity was measured based on the method described by McCleary and Sheehan (1989) and Barnes and Blakeney (1974). Briefly, 400 μl sample extracts were incubated at 37 °C for 30 min with phadebas tablets (Megazyme International Ireland Ltd., Bray, Ireland), mixed with a vortex and centrifuged at 960×*g* for 10 min, after which absorbance was measured with a spectrophotometer at 620 nm against a reagent blank and units were calculated in reference to a calibration curve.

One protease unit was defined as the amount of enzyme that releases 1.0 μg of phenolic compound, expressed as tyrosine equivalents, from a casein substrate per minute at pH 7.5 and 40 °C. Briefly, sample extracts were incubated at 50 °C for 60 min with a protazyme tablet (Megazyme International Ireland Ltd., Bray, Ireland) using pH 10 Tris/HCl buffers, centrifuged at 960×*g* for 10 min, after which the absorbance of supernatants was measured with a spectrophotometer at 590 nm against the blank and units were calculated in reference to a calibration curve.

## 2.4. Birds and management

At Massey University, day-old male broilers (Ross 308) were obtained, reared in floor pens and fed unsupplemented control diets prior to the introduction to cages, which were housed in an environmentally controlled room. The same procedures were followed at the University of Illinois using Ross 308 birds, except that the birds were kept in battery cages from the first day. Birds received a 20-h light programme and were allowed free access to feed and water. Birds were randomly allocated to cages and treatments at 12 d of age in trials 1 and 2 and at 7 d of age in trials 3 and 4. Experimental diets were fed from 12 to 21 d in trials 1 and 2 and from 7 to 21 d of in trials 3 and 4. The studies were conducted under approved animal use protocols that complied with the New Zealand Code of Practice for the Care and Use of Animals for Scientific Purposes (trials 1 and 2) and the Institutional Animal Care and Use Committee (trials 3 and 4), respectively.

## 2.5. $AME_n$ and AID measurements

In trials 1 and 2, $AME_n$ measurement was performed by total excreta collection. Feed intake and total excreta output were measured quantitatively per cage over four consecutive days starting at day 17. In trials 3 and 4, representative samples of excreta were collected daily on trays under each cage from 17 to 21 d of age. In all trials, excreta collections were pooled within a cage, mixed in a blender and sub-sampled. Each sub-sample was lyophilised, ground to pass through a 0.5 mm sieve and stored in airtight plastic containers at −4 °C pending analysis of dry matter, gross energy and nitrogen (N). Excreta samples from trials 3 and 4 were additionally analysed for titanium.

For assessment of AIDE and AID of AA and N, birds were euthanised by intracardial injection of sodium pentobarbitone in trials 1 and 2, or carbon dioxide asphyxiation in trials 3 and 4 and contents of the lower half of the ileum were obtained by gentle flushing with distilled water. The ileum was defined as the portion of the small intestine extending from the Meckel's diverticulum to a point approximately 40 mm proximal to the ileo-caecal junction (Ravindran et al., 2005). Digesta from birds within a cage were pooled and frozen immediately after collection, lyophilised and processed. Samples of digesta and diets were analysed for dry matter, N, AA, titanium and gross energy.

In trials 1 and 2, $AME_n$ and TTAR of N were calculated as reported by Ravindran et al. (2008). In trials 3 and 4, $AME_n$ and TTAR of N were calculated based on concentration of the indigestible marker in the diets using equations reported by Ravindran et al. (2005). Appropriate corrections were made for differences in moisture content. $AME_n$ was determined for zero N retention by multiplication with 34.4 kJ/g of N retained in the body (Hill and Anderson, 1958). Calculation of AID coefficients based on the concentration of indigestible marker was performed as reported by Ravindran et al. (2005).

**Table 2.** Expected and measured enzyme activities in samples of the experimental diets in four independent broiler digestibility trials.

| Trial | Xylanase[a] | | Amylase[b] | | Protease[c] | |
|---|---|---|---|---|---|---|
| | Expected | Measured | Expected | Measured | Expected | Measured |
| Trial 1 | | | | | | |
| Control | 0 | – | 0 | – | 0 | – |
| XA | 2000 | 2238 | 1800 | 2519 | 0 | – |
| PXA | 1000 | 1139 | 1800 | 2112 | 4000 | 3432 |
| Trial 2 | | | | | | |
| Control | 0 | – | 0 | – | 0 | – |
| XA | 2000 | 1800 | 1800 | 2011 | 0 | – |
| PXA | 1000 | 1056 | 1800 | 2108 | 4000 | 4482 |
| Trial 3 | | | | | | |
| Control | 0 | – | 0 | – | 0 | – |
| XA | 2000 | 1478 | 1800 | 1366 | 0 | – |
| PXA | 1000 | 941 | 1800 | 1505 | 4000 | 2899 |
| Trial 4 | | | | | | |
| Control | 0 | – | 0 | – | 0 | – |
| XA | 2000 | 1521 | 1800 | 2054 | 0 | – |
| PXA | 1000 | 912 | 1800 | 1771 | 4000 | 3597 |

[a] One xylanase unit was defined as the amount of enzyme that releases 0.48 μmol of reducing sugar as xylose from wheat arabino xylan per minute at pH 4.2 and 50 °C.

[b] One unit of amylase was defined as the amount of the enzyme catalysing the hydrolysis of 1 μmol glucosidic linkage per minute.

[c] One protease unit was defined as the amount of enzyme that releases 1.0 μg of phenolic compound, expressed as tyrosine equivalents, from a casein substrate per minute at pH 7.5 and 40 °C.

## 2.6. Chemical analysis

Dry matter content was determined using standard procedures (method 930.15; AOAC International, 2005). Gross energy was determined using adiabatic bomb calorimeter standardised with benzoic acid in all trials. N content was determined by the combustion method (method 968.06; AOAC International, 2005) using a CNS-2000 carbon, N and sulphur analyser. In trials 1 and 2, titanium content was measured on a UV spectrophotometer following the method of Short et al. (1996). AA, including cysteine and methionine, were determined in a Waters ion-exchange high performance liquid chromatographic system, as described by Ravindran et al. (2009). In trials 3 and 4, AA were analysed using the methods of AOAC International (2005). The samples were hydrolysed under N with 6 N HCl for 24 h at 110 °C before analyses were performed by automated cation-exchange chromatography. Analysis of cysteine and methionine involved performic acid oxidation before hydrolysis. Titanium analyses were conducted using procedures described by Myers et al. (2004).

## 2.7. Statistical analysis

Results from each of the trials were analysed separately using an ANOVA (Proc Mixed;SAS, 1997) that included the fixed effect of dietary treatment. Results of the four trials were then combined in a single data set and analysed using a model (Proc Mixed; SAS, 1997) that included the main effects of dietary enzyme, maize-DDGS inclusion and trial site as well as all the two-way interactions. Due to the heterogeneity of variances between sites, a heterogeneous variance model by trial site was specified using the Variance Components structure of Proc Mixed (SAS, 1997), which maximised fit. Least-squares-mean separation was made through multiple *t*-tests. Significance was assessed at $P<0.05$.

Linear regressions were performed to calculate the slope of the effect of enzymes on AID of AA compared to the control diets as a function of the amount of undigested AA in the control diets, using the REG Procedure of SAS (1997). Regressions were carried out separately for each of the enzyme treatments using the change in AID of AA for each AA compared to the control diets as independent variable. This change was calculated as the difference between the mean digestible AA in response to the enzyme and the mean digestible AA of the control treatment expressed in g AA/kg feed. The ileal undigested AA, calculated as the dietary AA minus the mean ileal digested AA of the control treatments and expressed in g AA/kg feed, was the dependent variable. No intercept was considered under the assumption that the improvement of digestibility at zero undigested AA was zero.

# 3. RESULTS

Initially, differences in AIDE, AID of N, TTAR of N and $AME_n$ due to enzyme supplementation were analysed for each of the four trials separately (Table 3). In trial 1, PXA increased ($P<0.05$) the AID of N while both XA and PXA increased ($P<0.05$) the $AME_n$ compared to the control treatment. AIDE and TTAR of N were unaffected by the enzyme treatments. In trial 2, both XA and PXA increased ($P<0.05$) the TTAR of N compared to the control treatment, but had no effects on the AIDE and N and $AME_n$. In trial 3, PXA increased ($P<0.05$) the AIDE and TTAR of N; and both XA and PXA increased ($P<0.05$) $AME_n$ compared to the control treatment. Enzymes, however, had no effect on the AID of N. In trial 4, both XA and PXA increased ($P<0.05$) the AIDE and PXA increased the TTAR of N compared to the control treatment, whereas XA and PXA increased ($P<0.05$) $AME_n$ in a stepwise manner.

Given the similarities in the design and diets and the fact that diets without and with DDGS were used in each research site, combined analyses of the four trials were also performed, which increased the sensitivity of the analyses and allowed studying the interaction between enzyme and diets. Due to heterogeneity of variance between trial sites, a heterogeneous variance model by trial site was specified in the mixed model, which improved fit statistics.

**Table 3.** Effects of enzyme supplementation on apparent ileal digestible energy (AIDE), coefficient of apparent ileal digestibility (AID) of nitrogen (N), total tract apparent retention (TTAR) of N and N-corrected apparent metabolisable energy ($AME_n$) in young broiler chickens in four independent trials.a [and] b

| Trial | Enzyme[c] | Maize-DDGS inclusion (g/kg) | AIDE (MJ/kg dry matter) | AID of N | TTAR of N (g/kg dry matter intake) | $AME_n$ (MJ/kg dry matter) |
|---|---|---|---|---|---|---|
| Trial 1 | Control | 0 | 13.30 | 0.806$^{y}$ | 23.3 | 12.66$^{y}$ |
| | XA | 0 | 13.50 | 0.817$^{xy}$ | 23.5 | 12.78$^{x}$ |
| | PXA | 0 | 13.57 | 0.829$^{x}$ | 23.9 | 12.81$^{x}$ |
| | SEM | | 0.09 | 0.005 | 0.20 | 0.04 |
| | Probability | | 0.13 | 0.012 | 0.21 | 0.035 |
| Trial 2 | Control | 100 | 13.38 | 0.793 | 23.5$^{y}$ | 12.68 |
| | XA | 100 | 13.74 | 0.810 | 24.3$^{x}$ | 12.82 |
| | PXA | 100 | 13.79 | 0.812 | 24.4$^{x}$ | 12.81 |
| | SEM | | 0.14 | 0.008 | 0.26 | 0.05 |
| | Probability | | 0.11 | 0.18 | 0.045 | 0.16 |
| Trial 3 | Control | 0 | 13.39$^{y}$ | 0.822 | 18.8$^{y}$ | 12.45$^{y}$ |
| | XA | 0 | 13.64$^{xy}$ | 0.821 | 20.0$^{xy}$ | 12.98$^{x}$ |
| | PXA | 0 | 13.73$^{x}$ | 0.833 | 20.4$^{x}$ | 13.17$^{x}$ |
| | SEM | | 0.11 | 0.007 | 0.45 | 0.08 |
| | Probability | | 0.10 | 0.40 | 0.06 | <0.001 |
| Trial 4 | Control | 70 | 13.11$^{y}$ | 0.780 | 21.4$^{y}$ | 12.66$^{z}$ |
| | XA | 70 | 13.59$^{x}$ | 0.784 | 20.9$^{y}$ | 13.01$^{y}$ |
| | PXA | 70 | 13.73$^{x}$ | 0.800 | 22.3$^{x}$ | 13.22$^{x}$ |
| | SEM | | 0.16 | 0.011 | 0.25 | 0.07 |
| | Probability | | 0.031 | 0.43 | 0.002 | <0.001 |

$^{x,y,z}$Means with no common superscripts within column and sub-grouping are different at $P<0.05$.

[a] Trials 1 and 2 consisted of six replicates per treatment, with six birds per replicate cage. Trials 3 and 4 consisted of eight replicates per treatment, with five birds per replicate cage.

[b] Ileal digesta or excreta samples were pooled by cage. Cage was considered as experimental unit.

[c] Negative control diets were supplemented with an enzyme complex containing xylanase and amylase (XA), or one containing protease, xylanase and amylase (PXA).

AIDE increased ($P<0.05$) with both enzyme combinations compared to the control diets (Table 4). The XA combination increased AIDE by 0.33 MJ/kg, whereas the PXA combination increased it by 0.41 MJ/kg compared to the control diet. AIDE of diets without and with maize-DDGS was similar ($P>0.05$). No interaction was evident between enzyme addition and maize-DDGS inclusion on the AIDE. The $AME_n$ followed a similar trend compared to that of AIDE, but there was a difference ($P<0.05$) between XA and PXA. XA increased $AME_n$ by 0.28 MJ/kg, while PXA increased $AME_n$ by 0.39 MJ/kg compared to the control diets. Increments of $AME_n$ due to enzymes were similar in diets without or with maize-DDGS, as indicated by the lack of enzyme×maize-DDGS interaction. TTAR of N increased ($P<0.05$) with PXA supplementation *versus* the control treatment ( Table 4), whereas XA supplementation did not cause a significant change from the control diet. There were significant effects of trial site on TTAR of N and AMEn, which were expected due to differences in ingredient sources. No significant interactions of enzyme and site were evident for AIDE and TTAR of N, but this interaction was significant for $AME_n$. The effect of XA and PXA on $AME_n$ compared to the control diets differed only in one of the sites and not in the other.

Significant ($P<0.05$–$0.001$) effects of enzymes on AID were evident for all analysed AA (Table 5). However, XA did not exhibit significant differences on the digestibility of any of the evaluated AA compared to the control diets, whereas PXA increased ($P<0.05$) the AID of N and all AA with the exception of methionine. The average increment in the AID of AA from the addition of PXA combination was 2.8%. There were significant effects of trial site on AID of AA for seven of the 17 analysed AA. Generally, no interactions between enzyme and trial site were present with the exception of asparagine, serine and threonine.

**Table 4.** Effects of enzyme supplementation and maize-DDGS inclusion on apparent ileal digestible energy (AIDE), total tract apparent retention (TTAR) of nitrogen (N) and N-corrected apparent metabolisable energy ($AME_n$) in young broilers.[a,b,c].

| Enzyme[d] | Maize-DDGS inclusion[e] (g/kg) | AIDE (MJ/kg dry matter) | TTAR of N (g/kg dry matter intake) | $AME_n$ (MJ/kg dry matter) |
|---|---|---|---|---|
| Control | | $13.29^{y}$ | $21.8^{y}$ | $12.61^{z}$ |
| XA | | $13.61^{x}$ | $22.2^{y}$ | $12.89^{y}$ |
| PXA | | $13.70^{x}$ | $22.8^{x}$ | $13.00^{x}$ |
| SEM | | 0.015 | 0.16 | 0.007 |
| | 0 | 13.51 | $21.7^{y}$ | 12.80 |
| | 70-100 | 13.55 | $22.8^{x}$ | 12.86 |
| | SEM | 0.012 | 0.13 | 0.006 |
| Effect[f] | | Probability | | |
| | | AIDE | TTAR of N | $AME_n$ |
| Enzyme | | <0.001 | 0.001 | <0.001 |
| Maize-DDGS | | 0.61 | <0.001 | 0.11 |
| Enzyme×maize-DDGS | | 0.47 | 0.97 | 0.74 |
| Site | | 0.84 | <0.001 | <0.001 |
| Site×enzyme | | 0.73 | 0.28 | <0.001 |
| Site×maize-DDGS | | 0.05 | <0.001 | 0.27 |

[x,y,z]Means with no common superscripts within column and sub-grouping are different at $P<0.05$.

[a] Data from four independent trials with 21-d old broiler chickens, two without and two with dietary inclusion of maize-DDGS, were pooled.

[b] Trials 1 and 2 consisted of six replicates per treatment, with six birds per replicate cage. Trials 3 and 4 consisted of eight replicates per treatment, with five birds per replicate cage.

[c] Ileal digesta or excreta samples were pooled by cage. Cage was considered as experimental unit.

[d] Negative control diets were supplemented with an enzyme complex containing xylanase and amylase (XA), or one containing protease, xylanase and amylase (PXA).

[e] Diets from trials 1 and 3 did not contain maize-DDGS, whereas trial 2 used 100 g/kg and trial 4 used 70 g/kg maize-DDGS.

[f] A heterogeneous variance model by site was specified using the Variance Components structure of Proc Mixed (SAS, 1997).

# 4. DISCUSSION

Analyses of each of the four trials separately demonstrated more consistent effects of enzymes at the total tract level than the ileal level, both for $AME_n$ and TTAR of N, although significant effects on AIDE and AID of N due to XA and PXA supplementation were present in one and two of the trials, respectively. This pattern may indicate that effects of these enzymes on nutrient digestibility were not limited to the small intestine, but the magnitude of the digestibility response increased during caecal digestion. Combined analyses of these four trials increased the statistical power and the ability to discern some of the main trends, in particular in terms of AID.

**Table 5.** Effects of enzyme supplementation and maize-DDGS inclusion on the apparent ileal digestibility (AID) coefficients of nitrogen (N) and amino acids in young broiler chickens.[a,b,c].

| | Enzyme[d] | | | | Maize-DDGS inclusion[e] (g/kg) | | | Probability[f] | | | | | |
|---|---|---|---|---|---|---|---|---|---|---|---|---|---|
| | Control | XA | PXA | SEM | 0 | 70–100 | SEM | Enzyme | Maize-DDGS | Enzyme × maize-DDGS | Site | Site × enzyme | Site × maize-DDGS |
| N | 0.800[y] | 0.808[xy] | 0.818[x] | 0.004 | 0.821[x] | 0.797[y] | 0.003 | 0.006 | <0.001 | 0.79 | 0.31 | 0.51 | 0.007 |
| Asparagine | 0.795[y] | 0.802[y] | 0.819[x] | 0.004 | 0.825[x] | 0.785[y] | 0.003 | <0.001 | <0.001 | 0.73 | 0.27 | 0.036 | 0.27 |
| Threonine | 0.733[y] | 0.739[y] | 0.765[x] | 0.005 | 0.765[x] | 0.726[y] | 0.004 | <0.001 | <0.001 | 0.68 | 0.030 | 0.024 | 0.08 |
| Serine | 0.792[y] | 0.792[y] | 0.816[x] | 0.004 | 0.813[x] | 0.787[y] | 0.003 | <0.001 | <0.001 | 0.47 | 0.06 | 0.003 | <0.001 |
| Glutamine | 0.868[y] | 0.872[y] | 0.885[x] | 0.003 | 0.885[x] | 0.865[y] | 0.003 | <0.001 | <0.001 | 0.98 | 0.32 | 0.16 | 0.09 |
| Proline | 0.818[y] | 0.825[y] | 0.840[x] | 0.004 | 0.833[x] | 0.823[y] | 0.003 | <0.001 | 0.022 | 0.94 | 0.08 | 0.41 | <0.001 |
| Glycine | 0.756[y] | 0.765[y] | 0.783[x] | 0.005 | 0.784[x] | 0.753[y] | 0.004 | <0.001 | <0.001 | 0.75 | 0.05 | 0.08 | 0.014 |
| Alanine | 0.821[y] | 0.825[y] | 0.843[x] | 0.004 | 0.838[x] | 0.821[y] | 0.004 | 0.001 | <0.001 | 0.86 | 0.97 | 0.17 | 0.043 |
| Valine | 0.792[y] | 0.801[y] | 0.818[x] | 0.005 | 0.819[x] | 0.788[y] | 0.004 | 0.001 | <0.001 | 0.98 | 0.58 | 0.29 | 0.08 |
| Isoleucine | 0.814[y] | 0.821[y] | 0.836[x] | 0.005 | 0.839[x] | 0.807[y] | 0.004 | 0.003 | <0.001 | 0.97 | 0.27 | 0.31 | 0.92 |
| Leucine | 0.837[y] | 0.840[y] | 0.860[x] | 0.004 | 0.851[x] | 0.840[y] | 0.003 | <0.001 | 0.017 | 0.92 | 0.13 | 0.35 | 0.07 |
| Tyrosine | 0.829[y] | 0.834[y] | 0.852[x] | 0.004 | 0.853[x] | 0.823[y] | 0.003 | <0.001 | <0.001 | 0.74 | 0.65 | 0.17 | 0.83 |
| Phenylalanine | 0.844[y] | 0.847[y] | 0.864[x] | 0.004 | 0.861[x] | 0.843[y] | 0.003 | <0.001 | <0.001 | 0.87 | 0.008 | 0.25 | 0.25 |
| Histidine | 0.821[y] | 0.824[y] | 0.842[x] | 0.004 | 0.839[x] | 0.819[y] | 0.003 | <0.001 | <0.001 | 0.88 | <0.001 | 0.14 | 0.08 |
| Lysine | 0.849[y] | 0.846[y] | 0.865[x] | 0.005 | 0.874[x] | 0.833[y] | 0.004 | 0.024 | <0.001 | 0.91 | 0.019 | 0.09 | 0.45 |
| Arginine | 0.870[y] | 0.873[y] | 0.889[x] | 0.004 | 0.891[x] | 0.864[y] | 0.003 | 0.003 | <0.001 | 0.81 | 0.001 | 0.09 | 0.59 |
| Cysteine | 0.705[y] | 0.714[y] | 0.743[x] | 0.006 | 0.740[x] | 0.702[y] | 0.005 | <0.001 | <0.001 | 0.70 | <0.001 | 0.22 | 0.005 |
| Methionine | 0.917[xy] | 0.914[y] | 0.926[x] | 0.003 | 0.934[x] | 0.904[y] | 0.003 | 0.036 | <0.001 | 0.98 | 0.003 | 0.34 | 0.42 |

[x,y,z]Means with no common superscripts within a column are different at $P<0.05$.

[a] Data from four independent trials with 21-d old broiler chickens, two with and two without dietary inclusion of maize-DDGS, were pooled.

[b] Trials 1 and 2 consisted of six replicates per treatment, with six birds per replicate cage. Trials 3 and 4 consisted of eight replicates per treatment, with five birds per replicate cage.

[c] Ileal digesta or excreta samples were pooled by cage. Cage was considered as experimental unit.

[d] Negative control diets were supplemented with an enzyme complex containing xylanase and amylase (XA), or one containing protease, xylanase and amylase (PXA).

[e] Diets from trials 1 and 3 did not contain maize-DDGS, whereas trial 2 used 100 g/kg and trial 4 used 70 g/kg maize-DDGS.

[f] A heterogeneous variance model by site was specified using the Variance Components structure of Proc Mixed (SAS, 1997).

Increments of energy digestibility in response to dietary carbohydrases without and with proteases have been previously reported in broilers fed maize-based diets (Meng and Slominski, 2005 and Cowieson and Ravindran, 2008), although some reports have not found effects in response to these enzyme combinations (Olukosi et al., 2007). Variability in energy responses to enzymes in poultry appears to depend on a number of factors, of which NSP structure and concentration are of greatest importance (Adeola and Cowieson, 2011). This study demonstrated increments in AIDE and $AME_n$ in maize-based diets in response to carbohydrases without or with protease. AIDE did not differ between XA and PXA fed broilers, but $AME_n$ differed by 0.11 MJ/kg, suggesting that the presence of protease in PXA increased the digestion in the hind gut of broilers compared to XA. The lack of interaction between enzymes and diets on energy digestibility and TTAR of N indicated that responses of both enzyme combinations were positive and in the same order of magnitude irrespective of the presence of maize-DDGS.

Improvements of AID of AA due to combinations of carbohydrases in poultry diets have been previously reported (Rutherfurd et al., 2007), although other reports have not found positive effects (Boguhn and Rodehutscord, 2010). It has been suggested that part of the improvements in AA digestibility due to carbohydrases are explained by a reduction of mucin secretion caused by the enhanced digestion of fibre and reductions in fibre components that irritate the gut lining (Cowieson and Bedford, 2009). Nonetheless, evidence to support that hypothesis in chickens is not conclusive (Sharma et al., 1997 and Fernandez et al., 2000). Amylase activity may also contribute to decreasing endogenous AA losses. Because starch is the major energy-yielding substrate in poultry diets, endogenous alpha-amylase secretions constitute an important part of the demands of AA and energy for digestion. In turn, pancreatic alpha-amylase expression has been demonstrated to be down-regulated as a feedback mechanism to the presence of exogenous amylase activity in the feed (Jiang et al., 2008). Reduced expression of amylase would ultimately result in increments of AID of AA and net energy.

Increments of AID of AA in broilers fed combinations of protease, xylanase and amylase have also been previously demonstrated (Cowieson and Ravindran, 2008). Nonetheless, the differential value of protease when applied in combination with xylanase and amylase had not been assessed. The current data indicate that the addition of protease was largely responsible for the observed increments in the coefficients of AID of AA. Even though XA did not exhibit significant effects on AA digestibility in the combined analysis of the four trials, it produced increments in AA digestibility in one of the trials (trial 1; data not shown). However, in the combined analyses, no evidence of interaction between trial site and enzymes was present with the exception of the AID of asparagine, threonine and serine. It may be hypothesised that the mechanism of action of the protease on AA digestibility is more dependent on the quality of protein in the diet, whereas effects of carbohydrases are more dependent on changes on endogenous secretions and the intestinal microbial environment. Although these aspects were not measured in the present studies, they might explain the variable responses to supplemental carbohydrases in different trials.

The reasons for the use of exogenous proteases in maize–SBM diets have been, first, to directly increase protein digestibility through the hydrolysis of dietary protein (Caine et al., 1998) and second, to disrupt protein–starch interactions in cereals, particularly maize (McAllister et al., 1993 and Belles et al., 2000), which would improve the digestibility of starch. Under the conditions of the present studies, the effect of protease on top of carbohydrase had a clear influence on AA digestibility in the small intestine. There was no effect on AIDE, but an effect on the $AME_n$ was present. These findings suggest that the direct hydrolysis of protein was more important in the small intestine and disruption of interactions of protein and other nutrients affected fermentation in the caecum. It appears, however, that post-ileal effects were not a function of the amount of NSP in the diet because there was no enzyme interaction with the presence of maize DDGS, which supplied additional insoluble NSPs.

The AA with the greatest digestibility response to PXA (Table 5) were cysteine (+5.4%), threonine (+4.4%), glycine (+3.6%) and valine (+3.3%). Threonine and cysteine have previously been reported as the most responsive to improvements in digestibility due to the PXA enzyme combination (Cowieson and Ravindran, 2008). These authors found improvements in AID of AA of +5.2% for threonine and +9.1% for cysteine when they included PXA on maize–SBM based diets. Similarly, Cowieson et al. (2006) reported increments on AID of +8.6% for threonine and +11.4% for cysteine with the inclusion of PXA and phytase in broilers fed maize–SBM diets. These AA, as well as valine, are particularly interesting to be included in matrix values of exogenous enzymes in diet formulations because they are limiting AA in maize–SBM broiler diets.

In contrast, the least responsive AA to PXA inclusion were methionine (+1.0), glutamine (+2.0), lysine (+1.9%) and arginine (+2.2%). Cowieson and Ravindran (2008) reported methionine (+0.5%), arginine (+0.8%) and glutamine (1.2%) as the AA with the lowest response to PXA, whereas lysine presented an intermediate response (+1.8%). Part of the low responsiveness of lysine and methionine in this trial may be explained by the inclusion of crystalline AA, which are highly digestible. The hydrolytic activity of exogenous proteases is not expected to affect crystalline AA. If it is assumed that proteolytic activity only affected the digestibility of the intact protein in the vegetable ingredients, the increment of AA digestibility of PXA would be underestimated in diets with inclusion of L-lysine. HCl and DL-methionine. Although the effect of PXA on lysine and methionine digestibility of the vegetable ingredients may be greater than it appears, this also means that improvements in digestibility of these two AA in practical diets are likely to be limited. Therefore, the effects of each of the most limiting essential AA should be considered independently when formulating diets supplemented with exogenous enzymes to avoid overestimating the contribution of the enzymes on the most limiting AA, particularly when they are supplemented with crystalline sources.

To further explore the reasons for the divergence in the digestibility response to PXA of different AA, the response relative to undigested fractions of each AA in the control diets was analysed (Fig. 1). It was previously suggested that the digestibility of nutrients is a major determinant of the digestibility responses to added exogenous enzymes in poultry diets (Cowieson, 2010). Analysing the data from 19 peer-reviewed studies, Cowieson (2010) reported a negative nonlinear relationship between improvements in the AID of AA due to supplemental xylanase and the AID of respective control diets, which suggests that a greater responses should be expected for AA with lower digestibility. Interestingly, the current study found that the best variable to explain the AA digestibility response to enzymes was not the digestibility coefficient, but the absolute amount of undigested AA of control diets at the ileal level. The amounts of ileal undigested AA were calculated by subtracting the apparent digestible AA (g/kg feed) of the control treatment from the AA concentrations of the control diet for each AA (g/kg feed).

Irrespective of the AA, a strong linear relationship between the amount of undigested AA and the response to enzymes was evident for PXA ($R^2$ = 0.94 and 0.96 for maize–SBM and maize–SBM–maize-DDGS diets, respectively; $P<0.0001$), but not for XA ($R^2$ = 0.42–0.59 for maize–SBM and maize–SBM–maize-DDGS diets, respectively; $P<0.01$). These data indicate that PXA increased the AID of the undigested AA fraction by 12–13%. The difference in the relationship for PXA *versus* XA supplementation suggests that protein hydrolysis catalysed by the exogenous protease was responsible for the improvements in digestibility of AA. This effect of the protease was non-specific among AA. It is important to recognise that the apparently undigested fraction contains both undigested AA of dietary origin, as well as endogenous secretions. However, if a reduction of endogenous AA losses had occurred, specific AA that are important constituents of mucin, *i.e.*threonine, serine, aspartic acid, glutamic acid and glycine ( Lien et al., 1996), would have increased in a greater proportion compared to the rest of AA in response to the PXA and XA treatments. It is possible that the effects of carbohydrases, *i.e.* xylanase, on mucin secretion, are less predictable because they depend on interactions with the gut wall and the intestinal microbiota. The direct effect of protease on dietary protein digestibility appears to be more predictable. This relationship did not significantly change in the diets tested in this study, although it is possible that the digestibility effects of these enzymes on the undigested fraction of other feed ingredients varies according to protein quality, which needs to be confirmed in future studies.

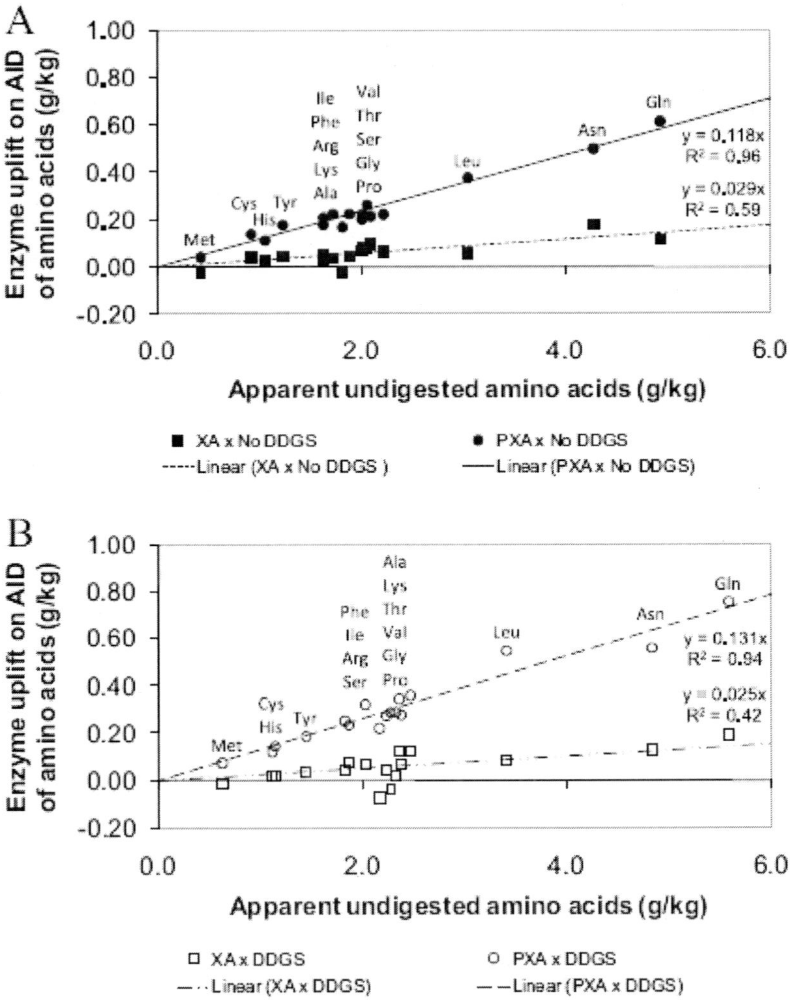


**Figure 1.** Change in apparent ileal digestibility (AID) of amino acids of 21-d broiler chickens in response to enzymes relative to the amounts of ileal undigested amino acids of the control diets without (A) or with (B) 7–10% maize-DDGS inclusion (XA = xylanase and amylase; PXA = protease, xylanase and amylase).

Results of the present studies were derived from broiler chickens fed mash diets, which may limit direct inferences to pelleted diets. Changes in chemical and structural characteristics of ingredients, particle size and chicken digestive development with pelleted diets (Amerah et al., 2007) may change the magnitude of the response of energy digestibility to different enzymes. Nonetheless, in reference to AA digestibility, pelleted diets may be expected to have a high potential for the action of exogenous enzymes, because poor gizzard development due to pelleting may not be beneficial to gastric digestion.

The present findings may allow the development of accurate AA matrices of exogenous enzymes for diet formulation, as long as reliable estimations of the undigested AA fraction in feed ingredients are available. Increments of energy and AA digestibility due to the use of exogenous carbohydrases and protease have the potential to reduce feed cost and contribute to reductions on the environmental footprint of poultry production systems.

## REFERENCES

1. Adeola, O., Cowieson, A.J., 2011. Opportunities and challenges in using enzymes to improve nonruminant animal production. J. Anim. Sci. 89, 3189–3218.

2. Amerah, A.M., Ravindran, V., Lentle, R.G., Thomas, D.G., 2007. Influence of feed particle size and feed form on the performance, energy utilization, digestive tract development and digesta parameters of broiler starters. Poult. Sci. 86, 2615–2623.

3. AOAC International, 2005. Official Methods of Analysis of AOAC. International, 18th ed. Assoc. Off. Anal. Chem., Arlington, VA. Barnes, W.C., Blakeney, A.B., 1974. Determination of cereal alpha-amylase. Starch 26, 193–197.

4. Belles, A.M., Montville, T.J., Wasserman, B.P., 2000. Enzymatic removal of zeins from the surface of maize starch granules. J. Ind. Microbiol. Biotechnol. 24, 71–74.

5. Boguhn, J., Rodehutscord, M., 2010. Effects of nonstarch polysaccharide-hydrolyzing enzymes on performance and amino acid digestibility in turkeys. Poult. Sci. 89, 505–513.

6. Caine, W.R., Verstegen, M.W.A., Sauer, W.C., Tamminga, S., Schulze, H., 1998. Effect of protease treatment of soybean meal on

content of total soluble matter and crude protein and level of soybean trypsin inhibitors. Anim. Feed Sci. Technol. 71, 177–183.

7. Choct, M., 2006. Enzymes for the feed industry: past, present and future. Worlds Poult. Sci. J. 62, 5–15.

8. Cowieson, A.J., 2005. Factors that affect the nutritional value of maize for broilers. Anim. Feed Sci. Technol. 119, 293–305.

9. Cowieson, A.J., Singh, D.N., Adeola, O., 2006. Prediction of ingredient quality and the effect of a combination of xylanase, amylase, protease and phytase in the diets of broiler chicks. 2. Energy and nutrient utilization. Br. Poult. Sci. 47, 490–500.

10. Cowieson, A.J., Ravindran, V., 2008. Effects of exogenous enzymes in maize-based diets varying in nutrient density for young broilers: growth performance and digestibility of energy, minerals and amino acids. Br. Poult. Sci. 49, 47-44.

11. Cowieson, A.J., Bedford, M.R., 2009. The effect of phytase and carbohydrase on ileal amino acid digestibility in monogastric diets: complementary mode of action? Worlds Poult. Sci. J. 65, 609–624.

12. Cowieson, A.J., 2010. Strategic selection of exogenous enzymes for maize/soy-based poultry diets. J. Poult. Sci. 47, 1–7.

13. Fernandez, F., Sharma, R., Hilton, M., Bedford, M.R., 2000. Diet influences the colonization of Campylobacter jejuni and distribution of mucin carbohydrates in the chick intestinal tract. Cell. Mol. Life Sci. 57, 1793–1801.

14. Gracia, M.I., Aranibar, M.J., Lazaro, R., Medel, P., Mateos, G.G., 2003. -Amylase supplementation of broiler diets based on corn. Poult. Sci. 82, 436–442.

15. Hill, F.W., Anderson, D.L., 1958. Comparison of metabolizable energy and productive energy determinations with growing chicks. J. Nutr. 64, 587–603.

16. Jiang, Z., Zhou, Y., Lu, F., Han, Z., Wang, T., 2008. Effects of different levels of supplementary alpha-amylase on digestive enzyme activities and pancreatic amylase mRNA expression of young broilers. Asian-Aust. J. Anim. Sci. 21, 97–102.

17. Lien, K.A., McBurney, M.I., Thomson, A.B.R., Sauer, W., 1996. Ileal recovery of nutrients and mucin in humans fed total enteral formulas supplemented with soy fiber. Am. J. Clin. Nutr. 63, 584–595.

18. McAllister, T.A., Phillippe, R.C., Rode, L.M., Cheng, K.J., 1993. Effect of the protein matrix on the digestion of cereal grains by ruminal microorganisms. J. Anim. Sci. 71, 202–212.

19. McCleary, B.V., Sheehan, H., 1989. Measurement of cereal - amylase: a new assay procedure. J. Cereal Sci. 6, 237–251.

20. Meng, X., Slominski, B.A., 2005. Nutritive values of corn, soybean meal, canola meal and peas for broiler chickens as affected by a multicarbohydrase preparation of cell wall degrading enzymes. Poult. Sci. 84, 1242–1251.

21. Myers, W.D., Ludden, P.A., Nayigihugu, V., Hess, B.W., 2004. Technical note: a procedure for the preparation and quantitative analysis of samples for titanium dioxide. J. Anim. Sci. 82, 179–183.

22. Olukosi, O.A., Cowieson, A.J., Adeola, O., 2007. Age-related influence of a cocktail of xylanase, amylase and protease or phytase individually or in combination in broilers. Poult. Sci. 86, 77–86.

23. Ravindran, V., Hew, L.I., Ravindran, G., Bryden, W.L., 2005. Apparent ileal digestibility of amino acids in dietary ingredients for broiler chickens. Anim. Sci. 81, 85–97.

24. Ravindran, V., Cowieson, A.J., Selle, P.H., 2008. Influence of dietary electrolyte balance and microbial phytase on growth performance, nutrient utilization and excreta quality of broiler chickens. Poult. Sci. 87, 677–688.

25. Ravindran, V., Morel, P.C.H., Rutherfurd, S.M., Thomas, D.V., 2009. Endogenous flow of amino acids in the avian ileum is increased by increasing dietary peptide concentrations. Br. J. Nutr. 101, 822–828.

26. Ritz, C.W., Hulet, R.M., Self, B.B., Denbow, D.M., 1995. Endogenous amylase levels and response to supplemental feed enzymes in male turkeys from hatch to eight weeks of age. Poult. Sci. 74, 1317–1322.

27. Rutherfurd, S.M., Chung, T.K., Moughan, P.J., 2007. The effect of a commercial enzyme preparation on apparent metabolizable energy, the true ileal amino acid digestibility and endogenous ileal lysine losses in broiler chickens. Poult. Sci. 86, 665–672.

28. SAS, 1997. SAS/STAT User's Guide Version 6.12. SAS Institute, Cary, NC. Sharma, R., Fernandez, F., Hilton, M., Schumacher, U., 1997. The influence of diet on the mucin carbohydrates in the chicken tract. Cell. Mol. Life Sci. 53, 935–942.

29. Short, F.J., Gorton, P., Wiseman, J., Boorman, K.N., 1996. Determination of titanium oxide added as an inert marker in chicken digestibility studies. Anim. Feed Sci. Technol. 59, 215–221.

# Index


## A

Alloxan model, 175
Anionic Diet, 94, 107
Apparent digestibility, 42, 244, 252

## B

Bacterial protein synthesis, 266
bioavailability, 203, 204, 205, 214
biochemistry, 204, 205, 206, 269
Broiler, 32, 293

## C

Canned pet food, 244
Carbohydrase, 293
Cats, 177, 182, 244, 247, 254, 260, 261
Crude protein, 266

## D

deficiency, 80, 81, 82, 165, 167, 179, 185, 204, 209, 210, 216, 232, 271
Dietary fiber, 260, 265, 266, 268, 274, 277, 285, 288

## F

Fatty liver, 68, 69
Feed additive, 2
Feed contamination, 144

## G

Gut health, 266
Gut microflora, 119, 128, 133, 137, 142

## H

hepatotoxic, 145, 222, 223, 228, 229, 230, 231, 232, 233, 235, 238
Hypomagnesaemia, 80, 103

## I

Immunity, 92, 144, 151
indigofera, 222, 236
indospicine, 221, 222, 223, 224, 225, 226, 227, 228, 229, 230, 231, 232, 233, 234, 236, 238
Intestinal morphology, 119

## K

Kafirin, 31, 32, 34, 58, 59
Ketosis, 70, 71, 88, 114, 184

## L

Laminitis, 73, 74, 105, 114

## M

Metritis, 84, 88, 114
Mycotoxins, 143, 144, 145, 154, 156, 163

## N

Nitrogen excretion, 266, 280
non-protein, 97, 222
Nutrient digestibility, 119, 124, 129, 133
Nutrition geometry, 32

## O

organic compounds, 204

## P

Paenibacillus polymyxa, 117, 118, 119, 120
Palm kernel cake, 119, 124, 127, 131, 133
Pig, 26, 144, 155, 259, 266, 281, 286
Probiotics, 1, 2, 3, 5, 6, 7, 8, 9, 10, 11, 13, 15, 24, 25, 26, 27, 141
Protease, 293

## R

Response surface design, 32
Rumen Acidosis, 72

## S

selenium, 40, 73, 204, 205, 206, 207, 208, 209, 210, 211, 215, 216, 217, 218, 219, 220, 249
selenoproteins, 203, 204, 206, 208, 214, 215, 216, 218
Sorghum, 32, 62, 64
Spray-dried plasma, 244, 249, 260
supplements, 3, 12, 73, 103, 204, 205, 209, 217, 286

## U

Udder Edema, 82, 107, 111, 113, 114

## V

Vaccine efficacy, 144

## W

Wet pet food, 244
Wheat gluten, 244